Coal in the 21st Century
Energy Needs, Chemicals and Environmental Controls

ISSUES IN ENVIRONMENTAL SCIENCE AND TECHNOLOGY

TITLES IN THE SERIES:

How to obtain future titles on publication:

A subscription is available for this series. This will bring delivery of each new volume immediately on publication and also provide you with online access to each title *via* the Internet. For further information visit http://www.rsc.org/issues or write to the address below.

For further information please contact:
Sales and Customer Care, Royal Society of Chemistry, Thomas Graham House, Science Park, Milton Road, Cambridge, CB4 0WF, UK
Telephone: +44 (0)1223 432360, Fax: +44 (0)1223 426017, Email: booksales@rsc.org
Visit our website at books.rsc.org

ISSUES IN ENVIRONMENTAL SCIENCE AND TECHNOLOGY

EDITORS: R.E. HESTER AND R.M. HARRISON

45
Coal in the 21st Century
Energy Needs, Chemicals and Environmental Controls

ROYAL SOCIETY
OF CHEMISTRY

Issues in Environmental Science and Technology No. 45

Print ISBN: 978-1-78262-860-6
PDF eISBN: 978-1-78801-011-5
EPUB eISBN: 978-1-78801-251-5
ISSN: 1350-7583

A catalogue record for this book is available from the British Library

The Royal Society of Chemistry is a charity, registered in England and Wales, Number 207890, and a company incorporated in England by Royal Charter (Registered No. RC000524), registered office: Burlington House, Piccadilly, London W1J 0BA, UK, Telephone: +44 (0) 207 4378 6556.

For further information see our web site at www.rsc.org

Preface

The long-term future for coal looks very bleak. The recent UN climate change conference (COP21) in Paris signaled an end to the fossil fuel era. And yet coal remains one of the world's most important sources of energy, fuelling over 40% of electricity generation worldwide. In many countries this figure is much higher: Poland relies on coal for over 94% of its electricity; South Africa for 92%; China for 77%; and Australia for 76%. India, the world's third largest energy consumer, generating over 70% of its electricity from coal in 2015, has set ambitious targets for increasing its coal-fired capacity over the next few years.

Coal has been the world's fastest growing energy source in recent years – faster than gas, oil, nuclear, hydro and renewables. During the next two decades several hundred million people worldwide will get electricity for the first time and, if current trends continue, most will use power produced by coal. It is essential also for steel and cement production and a host of other industrial activities. In China, coal is used on a grand scale for making many industrial chemicals, plastics, and liquid fuels – a role played by oil in most other countries.

However, the environmental impacts of the coal industry are hugely damaging. It is a major driver of climate change, almost 40% of the world's carbon dioxide emissions arising from burning coal. It kills thousands every year in mines and many more through air pollution. In presenting a critical review of coal in the 21st century, this book endeavours to provide a balanced view of both the positive and negative aspects of the topic, with an emphasis on the changing nature and future for the coal industry around the world.

Chapter 1, written by Liam McHugh of the World Coal Association, makes the case for coal being a critical enabler in the modern world, providing 41% of the world's electricity and being an essential raw material in the production of 70% of the world's steel and 90% of the world's cement. It has been and continues to be key to the building of modern societies and many countries have identified a continuing role for coal in their energy mix

Issues in Environmental Science and Technology No. 45
Coal in the 21st Century: Energy Needs, Chemicals and Environmental Controls
Edited by R.E. Hester and R.M. Harrison
© The Royal Society of Chemistry 2018
Published by the Royal Society of Chemistry, www.rsc.org

as they assess their future needs for access to energy, energy security and social and economic development. The 2015 'Paris Agreement' recognised that energy access and climate goals are not competing priorities but do require a balanced approach in order to achieve sustainable development.

Following on from this introductory overview, Zeshan Hyder of the Missouri University of Science and Technology reviews the methodology of coal mining in Chapter 2. Both surface and underground mining are described and the trade-offs in terms of economic benefits *versus* adverse environmental consequences are outlined. The world's largest producers of coal are China, United States, India, Australia and Indonesia.

Chapter 3 presents a technical examination of coal-fired power stations by Lucas Kruitwagen, Seth Collins and Ben Caldecott of the University of Oxford. The wide array of environment-related risks are detailed, including greenhouse gas emissions and stranded assets; water consumption and competition with agriculture, industry, and domestic uses; climate stresses induced by anthropogenic climate change; competition with renewables and generating flexibility; costs and trade-offs of mitigation options; retro-fitability with carbon capture and storage; and the availability of finance. The chapter concludes that the future of coal in the 21st century depends largely on the response of policy makers, industry and the concerned public to these risks.

In Chapter 4, Barbara Gottlieb and Alan Lockwood of the US organisation Physicians for Social Responsibility focus on the negative health impacts of coal, pointing out that coal-related pollution makes important contributions to four of the five leading causes of death in the US and that every aspect of coal use poses threats to health. Mining accidents, coal-worker's pneumoconiosis caused by sustained inhalation of coal dust, water and air pollution, toxic wastes, transportation and combustion pollution, greenhouse gas emissions and climate change all are examined critically as factors to be considered in assessing the cost of using coal.

Ken Kimmell and Rachel Cleetus of the Union of Concerned Scientists, which is based in Cambridge, Massachusetts, review the changing field of regulation in Chapter 5, providing insights on current policy and regulatory context for the major economies of China, India and Germany, as well as the United States. The transition to a low-carbon development is being driven largely by public health and climate change issues, but there are complex technological and socioeconomic questions to be addressed. This chapter provides a detailed overview of market trends and regulatory policy in these major coal-dependent countries, with particular emphasis on the US and the constraints of its political system.

In Chapter 6, Colin Snape of the University of Nottingham addresses the subject of liquid fuels and chemical feedstocks derived from coal. The direct liquefaction process involves catalysed reaction with hydrogen to cleave C–C bonds and remove heteroatoms (O, N and S), whereas indirect liquefaction utilises steam to transform coal into syngas, a mixture of carbon monoxide and hydrogen, which then can be converted catalytically into a wide variety of fuels and useful chemicals. This latter process has been the more widely utilised to date. The chapter focuses on the process conditions and the chemistry involved in converting coal to fuels and chemical feedstocks.

Finally, in Chapter 7, Ben Anthony of Cranfield University reviews the development and current status of carbon capture and storage (CCS). If the world is to meet its COP21 commitments and avoid ambient CO_2 levels in the environment reaching 450 ppm and above, limiting global temperature rise to 2 °C, CCS and/or related technologies appear to be essential. Renewable energy developments are unlikely to achieve this target on their own. However, although it is widely recognised that continuing large-scale use of fossil fuels worldwide does need to be accompanied by new technology to remove CO_2 from flue gases, the lack of political will to support the large-scale industrial deployment of CCS is telling. The alternative scenarios likely to result from global warming – forced population migration and famine – are alarming!

We are pleased to have been able to engage this international group of experts in producing a detailed and wide-ranging, well-balanced and informative review of coal in the 21st century. The importance of coal in meeting the world's energy needs as well as supporting major industries such as steel and cement production is unquestionable, but the environmental cost is high and there is an urgent need for mitigation strategies. The book provides crucial information and a balanced examination of both the positive and negative aspects of this controversial topic, which may be key to future policy-making at the highest level. It will additionally be of value to climate scientists and other scientists, engineers and managers involved in the energy and manufacturing industries and equally to students of environmental science and climate change.

Ronald E. Hester
Roy M. Harrison

Contents

Issues in Environmental Science and Technology No. 45
Coal in the 21st Century: Energy Needs, Chemicals and Environmental Controls
Edited by R.E. Hester and R.M. Harrison
© The Royal Society of Chemistry 2018
Published by the Royal Society of Chemistry, www.rsc.org

Coal Mining   30
Zeshan Hyder

Coal-fired Power Stations   58
Lucas Kruitwagen, Seth Collins and Ben Caldecott

Carbon Capture and Storage and Carbon Capture, Utilisation and Storage 198

E. J. "Ben" Anthony

Editors

Ronald E. Hester, BSc, DSc (London), PhD (Cornell), FRSC, CChem

Ronald E. Hester is now Emeritus Professor of Chemistry in the University of York. He was for short periods a research fellow in Cambridge and an assistant professor at Cornell before being appointed to a lectureship in chemistry in York in 1965. He was a full professor in York from 1983 to 2001. His more than 300 publications are mainly in the area of vibrational spectroscopy, latterly focusing on time-resolved studies of photoreaction intermediates and on biomolecular systems in solution. He is active in environmental chemistry and is a founder member and former chairman of the Environment Group of the Royal Society of Chemistry and editor of 'Industry and the Environment in Perspective' (RSC, 1983) and 'Understanding Our Environment' (RSC, 1986). As a member of the Council of the UK Science and Engineering Research Council and several of its sub-committees, panels and boards, he has been heavily involved in national science policy and administration. He was, from 1991 to 1993, a member of the UK Department of the Environment Advisory Committee on Hazardous Substances and from 1995 to 2000 was a member of the Publications and Information Board of the Royal Society of Chemistry.

Roy M. Harrison, BSc, PhD, DSc (Birmingham), FRSC, CChem, FRMetS, Hon MFPH, Hon FFOM, Hon MCIEH

Roy M. Harrison is Queen Elizabeth II Birmingham Centenary Professor of Environmental Health in the University of Birmingham. He was previously Lecturer in Environmental Sciences at the University of Lancaster and Reader and Director of the Institute of Aerosol Science at the University of Essex. His more than 500 publications are mainly in the field of environmental chemistry, although his current work includes studies of human health impacts of atmospheric pollutants as well as research into the chemistry of pollution phenomena. He is a past Chairman of the Environment Group of the Royal Society of Chemistry for whom he edited 'Pollution: Causes, Effects and Control' (RSC, 1983;

Fifth Edition 2014). He has also edited "An Introduction to Pollution Science", RSC, 2006 and "Principles of Environmental Chemistry", RSC, 2007. He has a close interest in scientific and policy aspects of air pollution, having been Chairman of the Department of Environment Quality of Urban Air Review Group and the DETR Atmospheric Particles Expert Group. He is currently a member of the DEFRA Air Quality Expert Group, the Department of Health Committee on the Medical Effects of Air Pollutants, and Committee on Toxicity.

List of Contributors

E. J. "Ben" Anthony, School of Water, Energy and Environment, Building 40, Cranfield University, Cranfield, Bedfordshire MK43 0AL, UK. Email: b.j.anthony@cranfield.ac.uk

Ben Caldecott, Smith School of Enterprise and the Environment, University of Oxford, South Parks Road, Oxford, OX1 3QY, UK. Email: ben.caldecott@smithschool.ox.ac.uk

Rachel Cleetus, Union of Concerned Scientists, Two Brattle Sq., Cambridge, MA 02138-3780, USA. Email: RCleetus@ucsusa.org

Seth Collins, Saïd Business School, University of Oxford, Park End Street, Oxford, OX1 1HP, UK. Email: Seth.Collins.mba2016@said.oxford.edu

Barbara Gottlieb, Physicians for Social Responsibility, 1111 14th Street NW, Suite 700, Washington, DC 20005, USA. Email: bgottlieb@psr.org

Zeshan Hyder, Missouri University of Science and Technology, Mining and Nuclear Engineering Department, Rolla, MO 65409, USA. Email: zhfy6@mst.edu

Ken Kimmell, Union of Concerned Scientists, Two Brattle Sq., Cambridge, MA 02138-3780, USA. Email: KKimmell@ucsusa.org

Lucas Kruitwagen, Smith School of Enterprise and the Environment, University of Oxford, South Parks Road, Oxford, OX1 3QY, UK. Email: Lucas.kruitwagen@smithschool.ox.ac.uk

Alan Lockwood, University at Buffalo, Buffalo, NY, and Physicians for Social Responsibility. Email: ahl@buffalo.edu

Liam McHugh, World Coal Association, Regent Street, London, W1 4JD, UK. Email: LMchugh@worldcoal.org

Colin E. Snape, University of Nottingham, Faculty of Engineering, Energy Technologies Building, Triumph Road, Nottingham, NG7 2TU, UK. Email: colin.snape@nottingham.ac.uk

World Energy Needs: A Role for Coal in the Energy Mix

LIAM McHUGH

ABSTRACT

The last 18 months have been a landmark period for climate, environment and development negotiation processes with the delivery of the Sustainable Development Goals and the Paris Agreement at the 21st Session of the Conference of the Parties (COP21). Energy access and climate goals are not competing priorities. As demonstrated by the Intended Nationally Determined Contributions (INDC) submitted in the lead-up to the Paris summit, each nation will choose an energy mix that best meets its needs. For this reason many countries have identified a continuing role for coal. Coal is a critical enabler in the modern world. It provides 41% of the world's electricity and is an essential raw material in the production of 70% of the world's steel and 90% of the world's cement. This chapter provides an introductory overview of coal, its use in building modern societies and its role in delivering low-cost on-grid electricity while integrating environmental commitments.

1 Introduction

There is little doubt that the 'Paris Agreement' delivered at the 2015 Climate Change Conference represented a landmark accomplishment. Equally, the rapid endorsement by national governments and the private sector to implement the deal is a welcome development.

Issues in Environmental Science and Technology No. 45
Coal in the 21st Century: Energy Needs, Chemicals and Environmental Controls
Edited by R.E. Hester and R.M. Harrison
© The Royal Society of Chemistry 2018
Published by the Royal Society of Chemistry, www.rsc.org

Looking ahead, as the deal becomes a reality it is vitally important that its delivery integrates environmental imperatives with the aims of universal access to energy, energy security and social and economic development. Only by balancing these elements can the agreement produce emissions reductions consistent with its vision while maintaining legitimate economic development and poverty alleviation efforts.

Energy is vital to development. Access to affordable and reliable electricity is the foundation of prosperity in the modern world. The Paris Agreement, however, has given countries an added impetus to ensure that improving energy access is balanced with action on reducing emissions.

Energy access and climate goals are not competing priorities. This understanding formed the basis of national climate pledges that were submitted in the months prior to the Paris climate negotiations and ultimately provided the foundation for the Agreement. Known as the Nationally Determined Contributions (NDCs), these pledges will act as strategic roadmaps for countries' climate, energy and development priorities.

Twenty four countries, representing over 50% of the world's emissions, submitted Intended Nationally Determined Contribution (INDCs) that identified a continuing role for coal.[1] No doubt as the INDC process is formalised this figure will rise.

Coal is a critical enabler in the modern world. It provides 41% of the world's electricity and is an essential raw material in the production of 70% of the world's steel and 90% of the world's cement.[2] Fossil fuels today provide over 80% of the world's primary energy, a percentage not forecast to change significantly for decades to come.[3] With the use of coal projected to continue to grow over the coming decades, a low-emission technology pathway for coal is required to meet emissions targets.

This pathway begins with deployment of high-efficiency, low-emissions (HELE) power stations using technology that is available today. These facilities are being built rapidly and emit 25–33% less CO_2 and eliminate other emissions, such as oxides of sulfur and nitrogen and particulates.[4] Moreover, HELE technology represents significant progress on the pathway towards carbon capture, use and storage (CCUS), which will be vital to achieving global climate objectives.

This chapter provides a comprehensive overview of coal and the role it plays in supporting sustainable development. It covers how coal is formed, how it is mined, through to its use and the impact it has on our societies and natural environments. It describes coal's role as an energy source and how coal – along with other sources of energy – will be vital in meeting the world's rapidly growing development needs along a sustainable pathway.

2   What Is Coal?

2.1   Coal Formation

Coal is a combustible, black or brownish-black sedimentary, organic rock, which is composed mainly of carbon, hydrogen and oxygen.

At its most basic level, coal is the altered remains of prehistoric vegetation that was originally located in swamps and peat bogs. Like all living organisms, these plants stored energy from the sun through a process known as photosynthesis. Generally, as plants die this energy is released during decay. At times, however, interruption of the decay process through the build-up of silt and other sediments, combined with tectonic movements, buried vegetation to great depths. In turn, buried vegetation underwent chemical and physical changes as a result of millions of years of pressure and heat transforming it into coal.

The process of coal formation began 360 to 290 million years ago during the Carboniferous Period – also known as the first coal age.

Several factors, including temperature, pressure and age, determined the quality of each coal deposit. Peat, in the first instance, was converted into lignite or 'brown coal'. Over millennia, pressure and temperature combined to transform lignite coal to more energy-intensive 'sub-bituminous' coal. Further chemical and physical changes occurred until these coals became harder and blacker, forming the 'bituminous' or 'hard' coals. Under the right conditions, the progressive increase in the organic maturity continued, finally forming anthracite.

2.2 Coal Classification

The degree of change that coal undergoes as it matures from lignite to anthracite is known as 'coalification'. The process has an important bearing on the physical and chemical properties of coal and is referred to as the 'rank' of the coal. Ranking is determined by the degree of transformation of the original plant material to carbon and is illustrated in Figure 1.

Low-rank coals – lignite and sub-bituminous – are low in carbon but high in hydrogen and oxygen content, and therefore lower in energy content. Low-rank coals are typically softer, friable materials with a dull, earthy appearance.

High-rank coals are high in carbon and therefore in heat value, with conversely lower levels of hydrogen and oxygen. Generally, high-rank coal is harder, stronger and often has a black, vitreous lustre. At the top of the rank

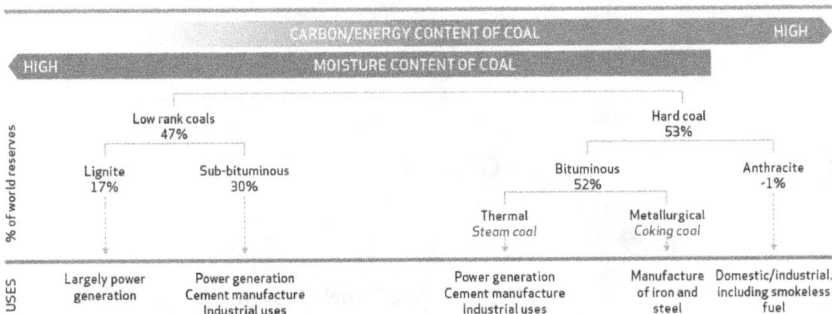

Figure 1 Types of coal.

scale, anthracite has the highest carbon and energy content, with the lowest levels of moisture.

2.3   Where Is Coal Found?

According to the International Energy Agency (IEA), there are over 985 billion tonnes of proven coal reserves worldwide.[5] At the current rate of production there are enough coal reserves to last around 128 years.[5]

Moreover, while it is estimated that current coal reserves could sustain demand well into next century, this could extend still further through a number of developments, including:

- The discovery of new reserves through ongoing and improved exploration activities
- Advances in mining techniques, which will allow previously inaccessible reserves to be reached.

As seen in Figure 2, coal reserves are available in almost every region, with recoverable reserves in almost 70 countries.[6] Although the largest reserves are in the USA, Russia, China and India, coal is actively mined in more than 70 countries. By contrast, Russia, Iran and Qatar control over half of the world's gas reserves and close to 50% of the world's oil reserves are located in the Middle East.[6]

Most coal is consumed domestically and only 17% is traded internationally.[7] In a number of countries coal is also the only domestically available energy

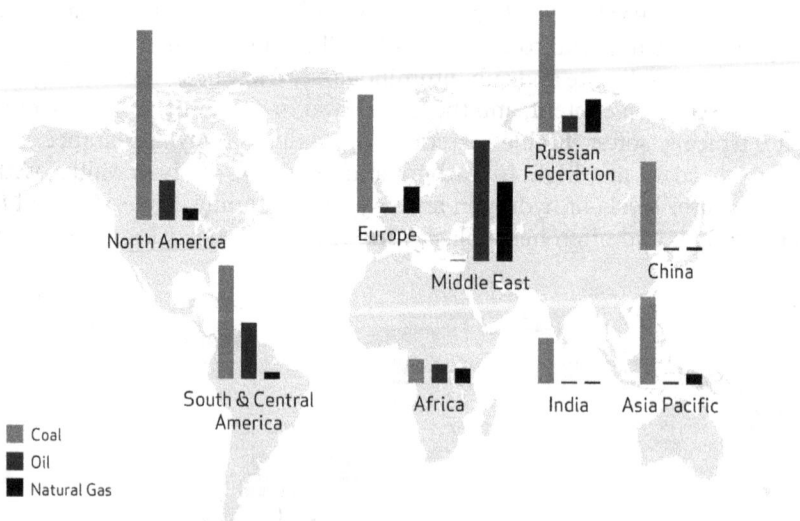

Source: BP 2015

Figure 2   Location of the world's main fossil fuels reserves (million tonnes of oil equivalent).

fuel and its use is motivated by both economic and energy security considerations. This is the case in countries and regions such as Europe, China and India, where coal reserves are much higher than oil or gas reserves. Most of the world's coal exports originate from countries that are considered to be politically stable – a characteristic which reduces the risks of supply interruptions.

2.4 Coal Exploration

Coal reserves are discovered through exploration activities. The process usually involves creating a geological map of the area, then carrying out geochemical and geophysical surveys, followed by exploration drilling. This allows an accurate picture of the area to be developed.

The area will only ever become a coal mine if it is large enough and of sufficient quality that the coal can be economically recovered. Once this has been confirmed, mining operations can begin.

2.5 Coal Mining

Coal is mined by two methods – surface (opencast) or underground (deep mining).

The choice of mining method is largely determined by the geology of the coal deposit. Underground mining currently accounts for about 60% of world coal production,[8] although in several important coal-producing countries surface mining is more common. Surface mining accounts for around 80% of production in Australia, while in the USA it is used for about 67% of production.[8]

2.5.1 Underground Mining. There are two main methods of underground mining: room and pillar and longwall mining.

In room and pillar mining, coal deposits are mined by cutting a network of 'rooms' into the coal seam and leaving behind 'pillars' of coal to support the roof of the mine. These pillars can be up to 40% of the total coal in the seam,[8] although this coal can sometimes be recovered at a later stage. This can be achieved in what is known as 'retreat mining', where coal is mined from the pillars as workers retreat. The roof is then allowed to collapse and the mine is abandoned.

Longwall mining involves the full extraction of coal from a section of the seam or 'face' using mechanical shearers. A longwall face requires careful planning to ensure favourable geology exists throughout the section before development work begins. The coal 'face' can vary in length from 100–350 m. Self-advancing, hydraulically-powered supports temporarily hold up the roof while coal is extracted. When coal has been extracted from the area, the roof is allowed to collapse. Over 75% of the coal in the deposit can be extracted from panels of coal that can extend 3 km through the coal seam.

The main advantage of room and pillar mining over longwall mining is that it allows coal production to start much more quickly, using mobile machinery that costs under $5 million (longwall mining machinery can cost $50 million).

The choice of mining technique is site-specific but always based on economic considerations; differences even within a single mine can lead to both methods being used.

2.5.2 Surface Mining. Surface mining – also known as opencast or opencut mining – is only economic when the coal seam is near the surface. This method recovers a higher proportion of the coal deposit than underground mining as all coal seams are exploited – 90% or more of the coal can be recovered.[8] Large opencast mines can cover an area of many square kilometres and use very large pieces of equipment, including: draglines which remove the overburden; power shovels; large trucks which transport overburden and coal; bucket wheel excavators; and conveyors.

The overburden of soil and rock is first broken up by explosives; it is then removed by draglines or by shovel and truck. Once the coal seam is exposed, it is drilled, fractured and systematically mined in strips. The coal is then loaded on to large trucks or conveyors for transport to either the coal preparation plant or direct to where it will be used.

2.6   Coal Preparation

Coal straight from the ground, known as run-of-mine (ROM) coal, often contains unwanted impurities such as rock and dirt and comes in a mixture of different-sized fragments. However, coal users need coal of a consistent quality. Coal preparation – also known as coal beneficiation or coal washing – refers to the treatment of ROM coal to ensure a consistent quality and to enhance its suitability for particular end-uses.[8]

The treatment depends on the properties of the coal and its intended use. It may require only simple crushing or it may need to go through a complex treatment process to reduce impurities. To remove impurities, the raw run-of-mine coal is crushed and then separated into various size fractions. Larger material is usually treated using 'dense medium separation'. In this process, the coal is separated from other impurities by being floated in a tank containing a liquid of high specific gravity, usually a suspension of finely ground magnetite. As the coal is lighter, it floats and can be separated off, while heavier rock and other impurities sink and are removed as waste.[8]

The smaller size fractions are treated in a number of ways, usually based on differences in mass, such as in centrifuges. A centrifuge is a machine which turns a container around very quickly, causing solids and liquids inside it to separate. Alternative methods use the different surface properties of coal and waste. In 'froth flotation', coal particles are removed in a froth produced by blowing air into a water bath containing chemical reagents. The bubbles attract the coal but not the waste and are skimmed off to recover the coal fines. Recent technological developments have helped increase the recovery of ultra-fine coal material.[8]

2.7   Coal Transportation

The way that coal is transported to where it will be used depends on the distance to be covered. Coal is generally transported by conveyor or truck over short distances. Trains and barges are used for longer distances within domestic markets, or alternatively coal can be mixed with water to form a coal slurry and transported through a pipeline.

Ships are commonly used for international transportation, in sizes ranging from Handymax (40–60 000 deadweight tonnage or DWT), Panamax (approximately 60–80 000 DWT) to large Capesize vessels (>80 000 DWT).[9] Around 1311 million tonnes (Mt) of coal was traded internationally in 2015 and around 90% of this was seaborne trade.[10] Coal transportation can be very expensive – in some instances it accounts for up to 70% of the delivered cost of coal.[9]

2.8   Coal Mining and the Environment

Measures are taken at every stage of coal transportation and storage to minimise environmental impacts.

Coal mining – particularly surface mining – requires large areas of land to be temporarily disturbed. This raises a number of environmental challenges, including soil erosion, dust, noise and water pollution, and impacts on local biodiversity. Steps are taken in modern mining operations to minimise these impacts. Good planning and environmental management minimises the impact of mining on the environment and helps to preserve biodiversity.[11]

2.8.1   Land Disturbance. In best practice, studies of the immediate environment are carried out several years before a coal mine opens in order to define the existing conditions and to identify sensitivities and potential problems. The studies look at the impact of mining on surface and ground water, soils, local land use, and native vegetation and wildlife populations. Computer simulations can be undertaken to model impacts on the local environment. The findings are then reviewed as part of the process leading to the award of a mining permit by the relevant government authorities.[12]

2.8.2   Mine Subsidence. A problem that can be associated with underground coal mining is subsidence, whereby the ground level lowers as a result of coal having been mined beneath. Any land use activity that could place public or private property or valuable landscapes at risk is clearly a concern. A thorough understanding of subsistence patterns in a particular region allows the effects of underground mining on the surface to be quantified. This ensures the safe, maximum recovery of a coal resource, while providing protection to other land uses.[11]

2.8.3 Water Pollution. Acid mine drainage (AMD) is metal-rich water formed from the chemical reaction between water and rocks containing sulfur-bearing minerals. The runoff formed is usually acidic and frequently comes from areas where ore- or coal-mining activities have exposed rocks containing pyrite, a sulfur-bearing mineral. However, metal-rich drainage can also occur in mineralised areas that have not been mined. AMD is formed when the pyrite reacts with air and water to form sulfuric acid and dissolved iron. This acid runoff dissolves heavy metals such as copper, lead and mercury into ground and surface water. There are mine management methods that can minimise the problem of AMD, and effective mine design can keep water away from acid-generating materials and help prevent AMD occurring. AMD can be treated actively or passively. Active treatment involves installing a water treatment plant, where the AMD is first dosed with lime to neutralise the acid and then passed through settling tanks to remove the sediment and particulate metals. Passive treatment aims to develop a self-operating system that can treat the effluent without constant human intervention.[11]

2.8.4 Dust and Noise Pollution. During mining operations, the impact of air and noise pollution on workers and local communities can be minimised by modern mine planning techniques and specialised equip-ment. Dust at mining operations can be caused by trucks being driven on unsealed roads, coal crushing operations, drilling operations and wind blowing over areas disturbed by mining. Dust levels can be controlled by spraying water on roads, stockpiles and conveyors. Other steps can also be taken, including fitting drills with dust collection systems and pur-chasing additional land surrounding the mine to act as a buffer zone between the mine and its neighbours. Trees planted in these buffer zones can also minimise the visual impact of mining operations on local communities. Noise can be controlled through the careful selection of equipment and insulation and sound enclosures around machinery. In best practice, each site has noise and vibration monitoring equipment in-stalled, so that noise levels can be measured to ensure the mine is within specified limits.[13]

2.9   Mine Rehabilitation

Coal mining is only a temporary use of land, so it is vital that rehabilitation of land takes place once mining operations have stopped. In best practice a detailed rehabilitation or reclamation plan is designed and approved for each coal mine, covering the period from the start of operations until well after mining has finished.[11]

2.10 Mining Safety

The coal industry takes the issue of safety very seriously. Coal mining deep underground involves a higher safety risk than coal mined in opencast pits. However, modern coal mines have rigorous safety procedures, health and safety standards and worker education and training, which have led to significant improvements in safety levels in both underground and opencast mining.[14]

Further detail on mining is given in Chapter 2 of this book.

3   The Global Coal Market

Coal is used by a variety of sectors – including power generation, iron and steel production, cement manufacturing and as a liquid fuel. The majority of coal is either utilised in power generation (steam coal or lignite) or in iron and steel production (coking coal).

3.1 Coal Production

Over 7700 Mt of coal is currently produced worldwide – a 70% increase over the past 20 years.[15] Coal production has grown fastest in Asia, while Europe has actually seen a decline in production. The largest coal-producing countries are not confined to one region – the top five producers are China, the USA, India, Australia and Indonesia.[16]

Global coal production is expected to increase by a further 4% through to 2040 – with developing Asia accounting for the vast majority of increased demand over this period.[20] Steam coal production is projected to have reached around 4.8 billion tonnes, with coking coal reaching 861 million tonnes.[17]

3.2 Coal Consumption

Coal plays a vital role in power generation and this role is set to continue. Coal currently fuels 41% of the world's electricity and this proportion is expected to remain at similar levels over the next 30 years.[18]

The biggest market for coal is Asia, which currently accounts for 75% of global coal consumption,[19] although China is responsible for a significant proportion of this. Many countries do not have natural energy resources sufficient to cover their energy needs, and therefore need to import energy to help meet their requirements. Japan, Chinese Taipei and Korea, for example, import significant quantities of steam coal for electricity generation and coking coal for steel production.

It is not just a lack of indigenous coal supplies that prompts countries to import coal but also the importance of obtaining specific types of coal. Major coal producers such as China, the USA and India, for example, also import quantities of coal for quality and logistical reasons. Coal will continue to play a key role in the world's energy mix, with demand in certain regions set to grow rapidly. Growth in both the steam and coking coal markets will be strongest in developing Asian countries, where demand for electricity and the need for steel in construction, car production, and demands for household appliances will increase as incomes rise.

3.3   Coal Trade

Coal is traded all over the world, with coal shipped huge distances by sea to reach markets.

Over the last twenty years, seaborne trade in steam coal has increased on average by about 17% each year,[20] while seaborne coking-coal trade has increased by 3% a year. Overall international trade in coal reached 1158 Mt in 2015;[21] while this is a significant amount of coal it still only accounts for about 15% of total coal consumed.[21]

Transportation costs account for a large share of the total delivered price of coal, therefore international trade in steam coal is effectively divided into two regional markets – the Atlantic and the Pacific. The Atlantic market is made up of importing countries in Western Europe, notably the UK, Germany and Spain. The Pacific market consists of developing and OECD Asian importers, notably Japan, Korea and Chinese Taipei. The Pacific market currently accounts for about 73% of world steam coal trade.[22] Markets tend to overlap when coal prices are high and supplies plentiful. South Africa is a natural point of convergence between the two markets.

4   How Is Coal Used?

Coal is a critical enabler in the modern world. It provides 41% of the world's electricity and is an essential raw material in the production of 70% of the world's steel and 90% of the world's cement.

4.1   Coal and Electricity

Coal-generated electricity is produced through the use of steam coal (also known as 'thermal coal'). Coal is first milled to a fine powder, which increases the surface area and allows it to burn more quickly. In these pulverised coal combustion (PCC) systems, the powdered coal is blown into the combustion chamber of a boiler where it is burnt at high temperature (see Figure 3). The hot gases and heat energy produced converts water – in tubes lining the boiler – into steam.[23]

The high-pressure steam is passed into a turbine containing thousands of propeller-like blades. The steam pushes these blades causing the turbine shaft to rotate at high speed. An electricity generator is mounted at one end of the turbine shaft. After passing through the turbine, the steam is condensed and returned to the boiler to be heated once again.[24]

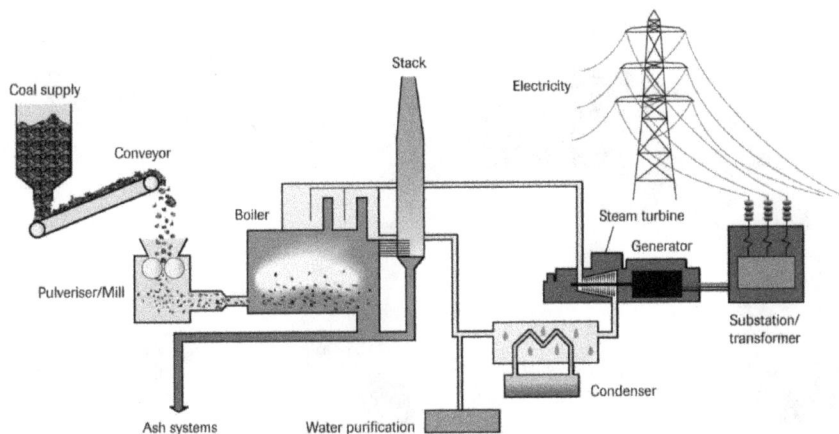

Figure 3   Converting coal to electricity.

The electricity generated is transformed into the high voltages (up to 400 kV) used for economic, efficient transmission *via* power line grids. When it nears the point of consumption, such as our homes, the electricity is transformed down to the safer 100–250 V systems used in the domestic market.[25]

Affordable, reliable and accessible energy is the foundation of prosperity in the modern world.

Since the 1980s, coal consumption has grown by over 140% in Brazil, 425% in India and 514% in China.[26] The economic and social progress made across these countries over the same period is well documented. China, in particular, has been a remarkable example of the role that affordable coal can play in improving access to energy and supporting economic development. Over the last three decades, according to World Bank estimates, 600 million people have been lifted out of poverty – almost all of those in China. Remove China from the mix and poverty levels in the rest of the world have barely improved. The link between access to affordable power from coal, economic growth and prosperity is clear. In China close to 99 percent of the population is connected to the grid.[27]

4.2   Coal's Role in Delivering Modern Infrastructure

Economic growth is dependent on the use of highly energy-intensive materials such as steel, cement, glass and aluminium. These materials are necessary for the construction and development of transport, energy, housing and water-management infrastructure.

4.3   Steel Production

Coal is a vital component of global steel production. One tonne of metal-lurgical coal is required to produce 1.3 tonnes of steel, equivalent to that

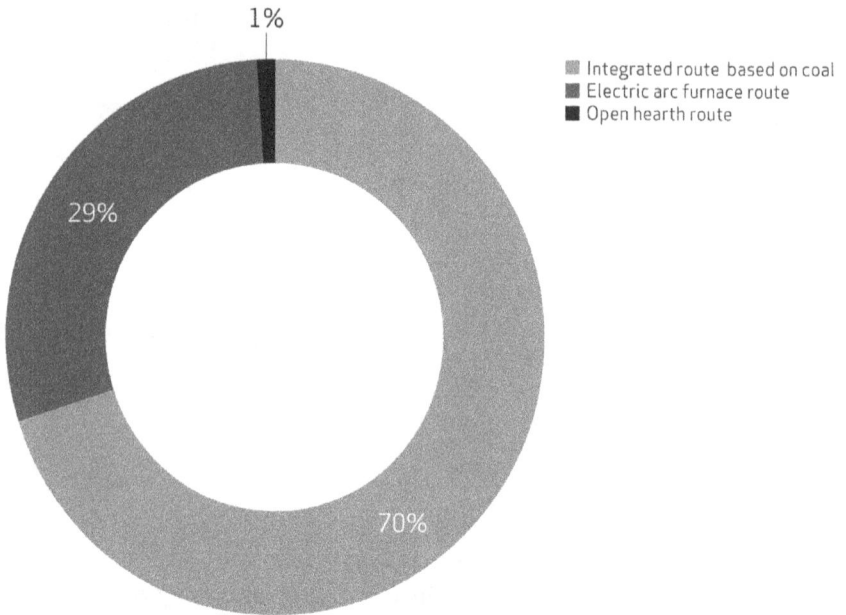

1%

- Integrated route based on coal
- Electric arc furnace route
- Open hearth route

29%

70%

Source: World Coal Association

Figure 4 Crude steel production by process.

needed to produce 18 household refrigerators.[28] Since 2001, average world steel use *per capita* has steadily increased from 150 kg in 2001 to 217 kg in 2014.[29] This is indicative of the rising living standards in the developing world and consequent demand for consumer goods.

There are two main steel production routes: the integrated steelmaking route and the electric arc furnace route. Coal is an essential raw material and energy fuel in both of them (see Figure 4). The integrated route, based on the blast furnace and basic oxygen furnace, uses on average 770 kg of coal, 1400 kg of iron ore, 150 kg of limestone and 120 kg of recycled steel to produce a tonne of crude steel. The electric arc furnace route uses, on average, 880 kg of recycled steel, 150 kg of coal and 43 kg of limestone to produce a tonne of crude steel.[30]

Steel production is critical in the construction of modern infrastructure such as transport, residential housing and commercial buildings. Manufacturing steel delivers the goods and services an advanced economy requires – healthcare, construction, telecommunications, improved agricultural practices, better transport networks, clean water and access to reliable and affordable energy.

4.4 Cement Production

Cement is the key ingredient in the production of concrete, an essential building material for society's infrastructure around the world, second only

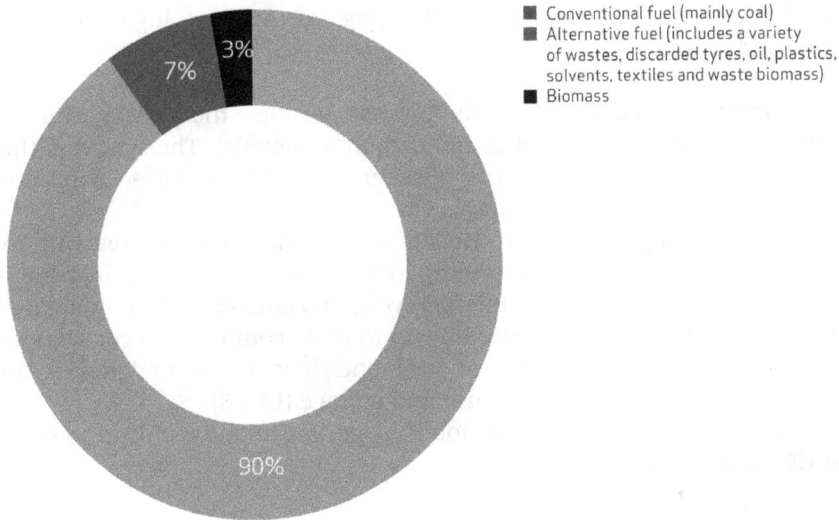

Conventional fuel (mainly coal)
Alternative fuel (includes a variety
of wastes, discarded tyres, oil, plastics,
solvents, textiles and waste biomass)
Biomass

Source: World Coal Association

Figure 5 Total fuel consumption by cement.

to water in total volumes consumed annually. Cement is essential for building houses, bridges, roads, dams, harbours and airports. Coal is used as an energy source in cement production to melt raw materials – limestone, silica, iron oxide and alumina (see Figure 5). It takes about 200 kg of coal to produce one tonne of cement and about 300–400 kg of cement is needed to produce one cubic metre of concrete.[31]

China provides an indication for the scope of cement (and therefore coal) demand that will be required over the coming decades. During the rapid period of industrialisation and urbanisation from 2011–2013, China produced more cement than the USA used in the entire 20th century.[32]

4.5 Coal Liquefaction

Coal-derived fuels, as well as coal-based electricity, can play a significant role in responding to the growing energy needs of the transport sector.

Liquid fuels from coal provide a viable alternative to conventional oil products and can be used in the existing supply infrastructure. Several coal-to-liquids (CTL) demonstration plants are being developed in China. CTL currently provides 20% of South Africa's transport needs, including 7.5% of jet fuel.[33]

Converting coal to a liquid fuel – a process referred to as 'coal liquefaction' – allows coal to be utilised as an alternative to oil. There are two different methods for converting coal into liquid fuels:

4.5.1 Direct Liquefaction. This works by dissolving the coal in a solvent at high temperature and pressure. This process is highly efficient, but the liquid products require further refining to achieve high-grade fuel characteristics.

4.5.2 Indirect Liquefaction. This process gasifies the coal to form a 'syngas' (a mixture of hydrogen and carbon monoxide). The syngas is then condensed over a catalyst – the 'Fischer–Tropsch' process – to produce high quality, ultra-clean products.

Coal-derived liquid fuels are also sulfur-free, low in particulates, with low levels of oxides of nitrogen, providing local and regional air quality benefits in comparison to oil. Over the full fuel cycle, CO_2 emissions from liquid fuels derived from coal can be reduced by up to 46%, compared to conventional oil products, if co-processing of coal and biomass is undertaken and combined with carbon capture, use and storage (CCUS).

Further detail on coal liquefaction is given in Chapter 6 of this book and on CCUS in Chapter 7.

4.6 Other Uses of Coal

Other important uses of coal include alumina refineries, paper manufacturers, and the chemical and pharmaceutical industries. Several chemical products can be produced from the by-products of coal. Refined coal tar is used in the manufacture of chemicals such as creosote oil, naphthalene, phenol and benzene. Ammonia gas recovered from coke ovens is used to manufacture ammonium salts, nitric acid and agricultural fertilisers. Thousands of different products have coal or coal by-products as components: soap, aspirins, solvents, dyes, plastics and fibres, such as rayon and nylon.

Coal is also an essential ingredient in the production of specialist products:

- Activated carbon – used in filters for water and air purification and in kidney dialysis machines.
- Carbon fibre – an extremely strong but light-weight reinforcement material used in construction, mountain bikes and tennis racquets.
- Silicon metal – used to produce silicones and silanes, which are in turn used to make lubricants, water repellents, resins, cosmetics, hair shampoos and toothpastes.

5 Meeting Future Energy Demand

One of the major challenges facing the world at present is that approximately 1.2 billion people – 16% of the world's population – live without any access to modern energy services.[34] A further 2.7 billion, or more than one in three people, use wood, charcoal, or animal waste for cooking and heating (see Figure 6).[35] In recent years, efforts to mobilise action in this area have

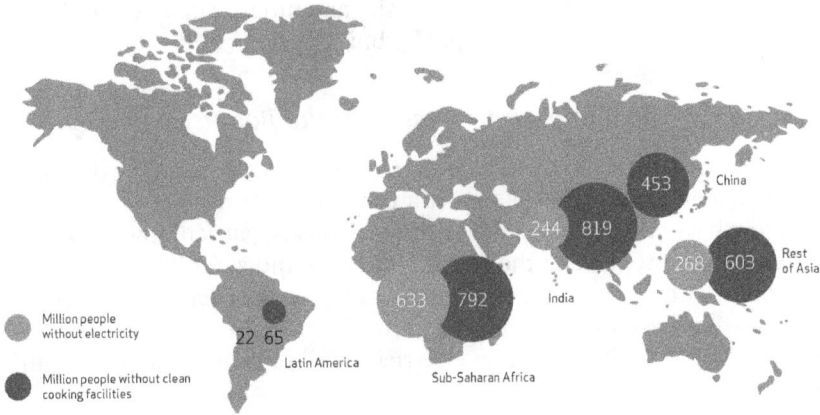

Source: International Energy Agency, World Energy Outlook 2016

Figure 6 People without access to modern energy services by region.

moved to the mainstream of development policy through initiatives such as the G20's Energy Access Action Plan, the UN's Sustainable Development Goals and the UN Secretary General's Sustainable Energy for All.

While progress to improve access to energy has been made in the recent past, quality of supply remains a significant challenge. The IEA attributes this to a variety of factors, including intermittent sources of energy in generation capacity. Such factors vary regionally, however, the implications are broadly the same: an economy that fails to operate at full potential.

Coal is a logical choice for providing universal, stable energy supplies in many countries impacted by energy poverty, as it is widely available, reliable and relatively low cost. It is for this reason that coal has become the *de facto* energy source for electrification. Research indicates that between 1990 and 2010 more than three quarters of a billion people – the vast majority of these being in developing countries – gained access to electricity due to coal-fired generation. The importance of coal for energy access is even clearer when compared with intermittent sources of power. Over the same twenty-year period, research finds that for every person who gained access to electricity from wind and solar, around 13 gained access due to coal.[36]

Coal use is forecast to rise over 11% through to 2040, with developing countries responsible for most of this increase to meet electrification rates.[37] Coal will therefore play a major role in supporting the development of base-load electricity where it is most needed, helping to bring economic growth to the developing world.

Further supporting this forecast, the IEA has projected that more than half of the on-grid electricity needed to meet their 'energy for all' scenario would need to come from coal.[38] It should be noted the even this scenario is not particularly ambitious, delivering an equivalent of five hours of electricity a day and excluding electricity for basic amenities, such as businesses, industry, hospitals, schools and public buildings.

5.1   Coal as an Important Element in the Balanced Energy Mix

The IEA, in its latest World Energy Outlook (2016), projects that global energy demand will grow by almost a third over the next 25 years. Most of this growth will take place in developing countries, particularly in South and Southeast Asia. At present, the average citizen of India, for instance, uses less than one tenth as much energy as the average OECD citizen.[39] Inevitably, development across the region will narrow this gap, necessitating a significant increase in their reliance on electricity and transportation and requiring major new supplies of energy.

To meet this need, the world cannot ignore any of the sources of energy available – including coal. Even in a world where renewables play a larger role in the energy mix, coal still has an important role to play. For example, in Germany, despite a surge in renewables, coal still provides base-load energy to ensure secure and reliable power supply.

It is therefore important that energy stakeholders pursue the most efficient pathway to ensure climate and energy goals. A balanced energy mix will make use of a variety of fossil fuels, nuclear and intermittent generation sources, with centralised and distributed models. Indonesia provides a clear example supporting this premise. The country has committed to adding 35 GW of additional electricity over the coming years to sustain economic growth and address electrification challenges.[40] Large, centralised coal plants will supply low-cost electricity to population and economic centres in Java, while off-grid renewables will be deployed for more remote parts of the archipelago. This approach, according to the World Economic Forum, will minimise expensive 'transmission and distribution construction, take advantage of local fuel sources and ensure affordable access to all'.

5.2   Coal as a Guarantor of Energy Security

The global coal market is large and diverse, with many different producers and consumers from every continent. Coal supplies do not come from one specific area, which would make consumers dependent on the security of supplies and stability of only one region. They are spread out worldwide and coal is traded internationally. There are enough coal reserves to last for around 110 years at the current rate of production. In comparison, proven oil and gas reserves are expected to last an equivalent of approximately 51 and 53 years, respectively, at current production levels.[41]

Many countries rely on domestic supplies of coal for their energy needs – such as China, the USA, India, Australia and South Africa. Others import coal from a variety of countries; in 2015, Germany, for example, imported coal from Russia, Colombia and the USA, as well as smaller amounts from a number of other countries and its own domestic supplies.

Coal therefore has an important role to play in maintaining the security of the global energy mix, complementing other fuels and energy sources that are generally more vulnerable to disruption. Coal contributes to security of the energy mix in a variety of ways:

- Coal reserves are very large and will be available for the foreseeable future without raising geopolitical or safety issues.
- Coal is readily available from a wide variety of sources in a well-supplied worldwide market.
- Coal can be easily stored at power stations and stocks can be drawn on in emergencies.
- Coal-based power is not dependent on the weather and can be used as a backup for wind and hydropower.
- Coal does not need high-pressure pipelines or dedicated supply routes.
- Coal supply routes do not need to be protected at enormous expense.

These features help facilitate efficient and competitive energy markets and help stabilise energy prices.

5.3   On-grid Electricity

In recent years, renewable energy has proven particularly adept at providing off-grid electricity for remote communities in the least-developed economies. Urbanising and industrialising economies, however, will require the development of on-grid base-load electricity. Renewables are an intermittent source of electricity – the wind doesn't blow all the time, the sun doesn't shine 24 hours a day. Base-load power is essential to support reliable, stable electricity grids. Fossil fuels, including coal are major providers of base-load power. Development of base-load infrastructure enables societies to provide greater modern lifestyle services, such as refrigeration and air-conditioning, but will also ensure reliable electricity for businesses, hospitals, public services and industry.

Indeed, this understanding was recently articulated by India's Power, Coal and Renewable Energy Minister, Piyush Goyal, who stated "We will be expanding our coal-based thermal power. That is our base-load power. All renewables are intermittent. Renewables have not provided base-load for anyone in the world".[42]

5.3.1   The Logical, Low-cost, Base-load Power Choice. The link between access to affordable power from coal, economic growth and prosperity is clear. In India, for example, electricity produced from coal-based power plants is 30% cheaper than electricity produced from renewables (and 16% cheaper than domestic natural gas).

Looking ahead, coal is likely to remain the most affordable fuel for power generation in many developing countries for decades to come. Analysis presented in the World Coal Association's (WCA's) 2016 study, *The Power of*

High Efficiency Coal, supports this forecast. The report considers the Levelised Cost Of Electricity (LCOE) for various technologies in non-OECD Asia. The LCOE provides an estimated price that generators are likely to incur producing electricity from various sources. The metric takes into account all of a system's expected lifetime costs (including construction, financing, fuel, maintenance, taxes, insurance and incentives), which are then divided by the system's lifetime expected power output (kWh). The calculation is vitally important for national power development plans to compare the costs of different technologies.

In 2035, data suggest that coal will generate electricity at a lower cost than other technologies – including gas – in The Association of Southeast Asian Nations (ASEAN) countries plus India, Bangladesh, Sri Lanka and Pakistan (grouped as South East Asia in Figure 7). The LCOE cost for the HELE technologies – supercritical coal (SC), ultra-supercritical coal (USC) and integrated gasification combined cycle (IGCC) – ranges from 55 to 60 $ MWh^{-1}. In comparison, the LCOE cost of Open Cycle Gas Turbine (OCGT) is almost double. It should also be taken into consideration that for many of these countries coal is more readily available than gas, which requires the development of pipeline infrastructure for its delivery. The comparative cost advantages of coal generation are even clearer in China – the main economy represented in the 'Rest of non-OECD Asia' in Figure 7. The various HELE technologies have an LCOE of around $50 MWh^{-1}, a third of the price of open-cycle gas turbines.[43]

5.3.2 Low-cost Electricity from Coal Aids Industrialisation. Stable and affordable energy pricing is a vital consideration for developing countries working to build industrial capacity, particularly in energy-intensive industries. Fluctuating fuel prices and high electricity costs can lead to a loss of competitive advantage and, in prolonged cases, loss of the industry altogether.

As noted earlier, electricity produced from coal-based power plants is 30% cheaper in India than electricity produced from renewables (and 16% cheaper than domestic natural gas). This has clear benefits for citizens through reduced energy costs, but the effect of lower energy prices are even clearer for business. Lower energy prices are a key factor in fostering industrial competitiveness. Lower cost electricity produced by coal results in lower production costs, increasing profits for industry (and thereby the country), and in turn promoting further economic activity.

The recent construction of the Sasan Power coal-fired power station in Madhya Pradesh, India, supports this premise. The facility has been credited with adding $2.5 billion to the local economy through improved energy access for local businesses. Improved energy access has allowed electric pumps to be deployed in the state's farming and allied services sector. Farmers have also benefited from improved access to market information, such as online agricultural market places. These developments have led to productivity gains through increased yields and more crop diversity.[44]

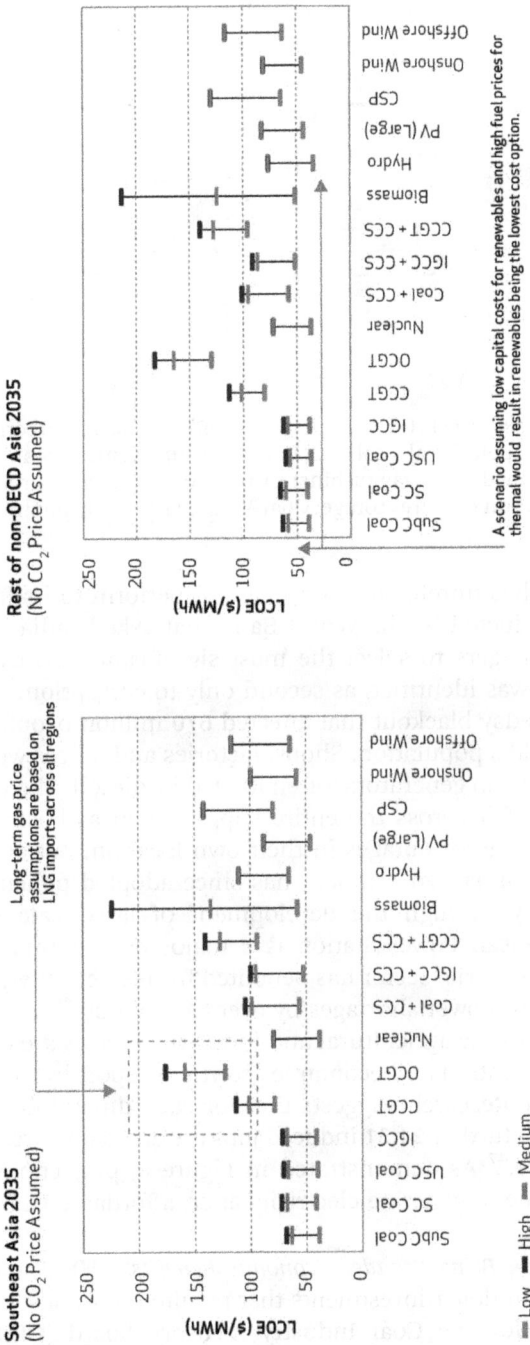

Figure 7 Lifetime cost of electricity *per* MWh across generation technologies in 2035. (Subcritical coal: SubCCoal; supercritical coal: SCCoal; ultra-supercritical coal: USC Coal; integrated gasification combined cycle: IGCC; open cycle gas turbine: OCGT; coal with carbon capture storage: Coal + CCS; large-scale photovoltaics: PV large; concentrated solar power: CSP).

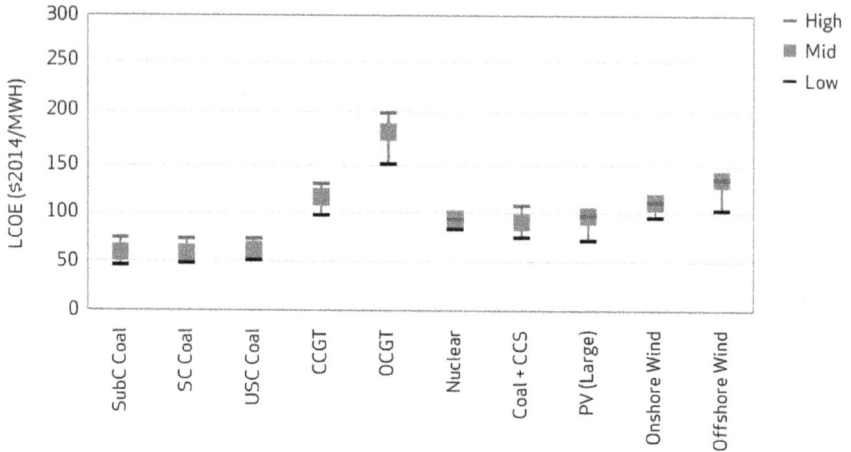

Figure 8   India Levelised Cost Of Electricity (LCOE) in 2035. (Subcritical coal: SubCCoal; supercritical coal: SCCoal; ultra-supercritical coal: USC Coal; IGCC; combined cycle gas turbine: CCGT; open cycle gas turbine: OCGT; coal with carbon capture storage: Coal + CCS; large-scale photovoltaics: PV large).

An economy that has unreliable energy cannot perform to its full potential. In a survey conducted by the World Bank that asked Indian business owners and top managers to select the most significant obstacle to their business, electricity was identified as second only to corruption.[45] In 2012, India endured a two-day blackout that affected 670 million people — more than 10% of the world's population. Shops, factories and offices were forced to close or rely on backup generators (often fuelled by diesel). The impact on daily commerce was felt across the entire supply chain as businesses not only had to compensate for outages in their own location, but also that of their suppliers. The Indian government has since adopted policies to promote energy security through the development of large-scale coal-fired power plants. The Sasan Power Station is a major part of this outcome. Madhya Pradesh's industrial sector has benefited from stable energy supply, reducing losses due to power shortages by over $1.5 billion.[46]

Productivity gains in the agricultural and industrial sectors are critical for a country's industrialisation and economic growth. A report by the National Council of Economic Research suggests that for each direct job created in the electricity sector a further 24.31 induced jobs are created in areas such as retail and agriculture.[47] As demonstrated in Figure 8, projections suggest that coal will continue to generate electricity at an affordable LCOE.

5.3.3   Coal Sector can Bring Broader Economic Benefits.   HELE coal power plants are multi-billion dollar investments that require thousands of workers. Analysis conducted for the Coal Industry Advisory Board (CIAB) found that during the four-year construction of a supercritical facility in India over 5000 workers were directly employed. The economic activity that construction of the facility bought created a further 3700 indirect jobs. Once operational, HELE plants have life-cycles of several decades and directly employ engineers of various disciplines, managers, and personnel related to finance, administration, human resources and security. Indirect job creation continues once HELE facilities become operational. The CIAB analysis found that close to 4000 jobs were created through local contracts for manpower supply, housekeeping, horticulture works and vehicle supply. Moreover, the report found that employment created by the facility stimulated economic activity across the local community. Those directly or indirectly employed at the plant tended to have higher disposable incomes and spend more on in items like food, consumer durables and leisure activities. According to this analysis, this created close to 30 000 jobs over the construction and operational phases of the facility's lifecycle.[47]

Furthermore, developing coal reserves for use within the local electricity sector provides benefits across the economy. While coal extraction methods vary considerably across regions, a common characteristic is that mining is a highly capital-intensive process that results in much economic activity.

In South Africa, one of the world's major producers and consumers, the coal-mining industry comprises more than 17% of the mining workforce – that is, more than 85 000 workers. The country's Department of Mines Resources estimates that the coal industry in South Africa pays more than US$1.2 billion in wages to its workforce each year.[48] While the skill required for positions within coal mining varies, much of the training – such as mechanical, technical, health and safety – is provided through employment in the sector.

Coal-mining activities are also an important economic contributor for governments. In 2014, South Africa earned more than US$3 billion in export revenue from coal – roughly one-quarter of total mining GDP.[48] The sector is also a leading contributor to government revenues through royalties, which according to the National Treasury Revenue Estimates totalled US$ 400 million in 2013.

In the recent past, development of coal mining has also bought co-benefits through infrastructure development. Rail lines and roads required to transport coal from the mine/port can be utilised by a variety of other industries. For instance, Mozambique's Nacala port, which began operations in 2015, was originally developed as an export terminal for the country's coal resources. In addition to enabling the export of coal, the port has allowed Mozambique to develop as a regional transportation hub. Improved logistics have benefited other neighbouring economies, particularly Mozambique's landlocked neighbour Malawi.[49]

6 Coal and the Environment

As the Paris Agreement is formally adopted, it is vitally important that its implementation integrates environmental imperatives with the aims of universal access to energy, energy security and social and economic

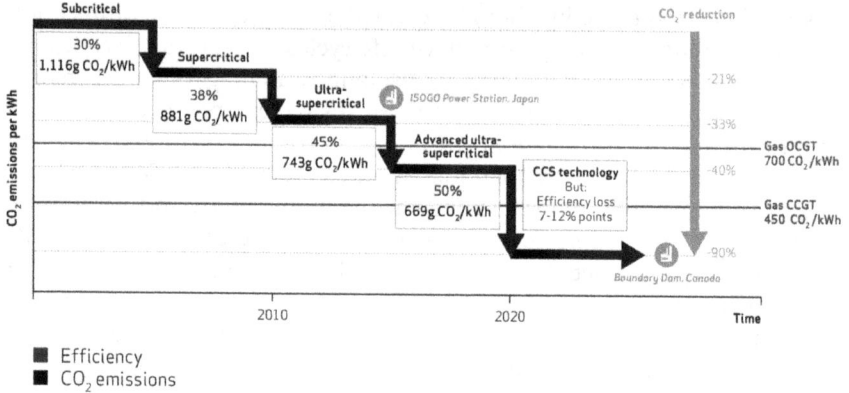

Figure contents (labels within figure):
- Subcritical — 30% — 1,116g CO_2/kWh
- Supercritical — 38% — 881g CO_2/kWh
- Ultra-supercritical — 45% — 743g CO_2/kWh
- ISOGO Power Station, Japan
- Advanced ultra-supercritical — 50% — 669g CO_2/kWh
- CCS technology — But: Efficiency loss 7-12% points
- CO_2 reduction — -21% — -33% — -40% — -90%
- Gas OCGT 700 CO_2/kWh
- Gas CCGT 450 CO_2/kWh
- Boundary Dam, Canada
- CO_2 emissions per kWh
- 2010 — 2020 — Time
- ■ Efficiency
- ■ CO_2 emissions
- Source: VGB PowerTech 2013

Figure 9 CO_2 reduction potential of coal-fired power plants by increased efficiency.

development. Only by balancing these elements can the agreement produce emissions reductions consistent with the vision of the Paris Agreement and with the broader development objectives.

Energy is an enabler of development. Access to affordable and reliable electricity is the foundation of prosperity in the modern world. As demonstrated by the Nationally Determined Contributions (NDCs), each nation will choose an energy mix that best meets its needs. For this reason many countries have identified a continuing role for coal. This is especially true for the rapidly urbanising and industrialising economies of Asia where coal is forecast to be an integral fuel source and vital to economic growth.

This begins with deployment of HELE power generation using technology that is available today. These facilities are being built rapidly and, as shown in Figure 9, emit 25–33% less CO_2 and significantly reduce or eliminate emissions, such as oxides of sulfur and nitrogen, and particulates.

Moreover, HELE technology represents significant progress on the pathway towards carbon capture, use and storage (CCUS), which will be vital to achieving global climate objectives.

Recognising the above, many countries have included a role for HELE coal-based power generation in their INDCs. In order to support the transition away from older, less efficient subcritical technology, these countries will require international financial, technological and other kinds of support to accelerate deployment of this technology.

6.1   Efficiency Improvements – What Can Be Achieved?

HELE technologies are in existence and available 'off-the-shelf'. They are currently being installed and used in many countries and provide efficiency gains and are financially viable. HELE coal-fuelled generation facilities, with modern emissions-control systems, emit 25–33% less CO_2 and significantly reduce or eliminate pollutant emissions, such as oxides of sulfur and nitrogen, and particulates compared to older, less-efficient subcritical technology.

The current average efficiency of coal-fired power plants around the world is 33%. This is well below the state-of-the-art rate of 45% and even 'off-the-shelf' rates of around 40%. Increasing the efficiency of coal-fired power plants by 1% reduces CO_2 emissions by between 2 and 3%. Moving the current average global efficiency rate of coal-fired power plants from 33% to 40% by deploying more advanced technology could cut 2 gigatonnes of CO_2 emissions.[50]

Recent trends suggest that the link between energy demand and economic growth is gradually becoming eroded. Indeed, the 2013 BP Energy Outlook forecasts energy intensity – the amount of energy required *per* unit of GDP – to decline by 36% (1.9% *p.a.*) between 2012 and 2035. This is a promising development and may in part be attributed to the number of countries that have committed to deploy HELE technologies, rather than older less-efficient subcritical coal-fired power stations.

Yet, there is a risk of complacency following the Paris Agreement. Over the coming decades, developing countries will have to balance environmental imperatives with the aims of universal access to energy, energy security and social and economic development. Without international financial, technological and other kinds of support to accelerate deployment of HELE technology, it is possible that the transition away from subcritical technologies will stall, weakening the recent gains in reducing energy intensity.

Analysis presented in the 2015 WCA study, *India's Energy Trilemma*, provides compelling evidence on the emission-reduction benefits of deploying HELE technologies. This analysis shows that replacing the subcritical capacity currently in India's development pipeline with supercritical or ultra-supercritical capacity could remove up to 11 billion tonnes of CO_2 emissions over the life of the power plants (see Figure 10). Adopting the most

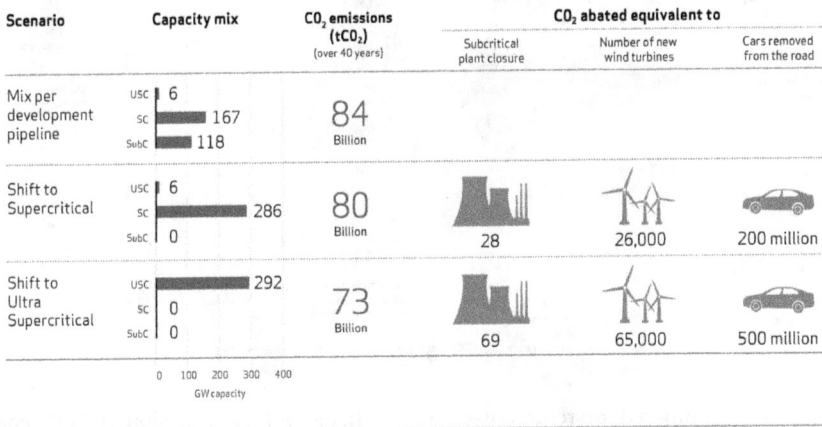

Scenario	Capacity mix		CO_2 emissions (tCO$_2$) (over 40 years)	CO_2 abated equivalent to		
				Subcritical plant closure	Number of new wind turbines	Cars removed from the road
Mix per development pipeline	USC	6	84 Billion			
	SC	167				
	SubC	118				
Shift to Supercritical	USC	6	80 Billion	28	26,000	200 million
	SC	286				
	SubC	0				
Shift to Ultra Supercritical	USC	292	73 Billion	69	65,000	500 million
	SC	0				
	SubC	0				

0   100   200   300   400
GW capacity

Figure 10   The environmental benefits of deploying cleaner coal technology in India.

efficient HELE technology would abate the CO_2 emissions equivalent to removing 500 million cars from the road. While the environmental implications are clear, the report notes that, with limited financing options available from development banks, power plant developers may accept lower efficiency and poorer emissions rates due to cost differences. There is as much as a 40% price difference between the capital costs of an ultra-supercritical and a subcritical coal plant. Analysis suggests that if all coal plants built from 2020 onwards were ultra-supercritical, total capital expenditure would reach $500 billion by 2040, compared to around $387 billion if all coal plant built from 2020 onwards were subcritical.[51]

Complementary analysis presented in the WCA's 2016 report, *The Power of High Efficiency Coal*, found that in many scenarios HELE technologies represent the lowest cost CO_2 abatement alternative (on a $ tonne^{-1} basis). The report considered the implications of concentrating support on renewable technology and excluding support for HELE coal-fired generation. The study found the challenge with increasing funding for the deployment of renewable generation is that its low load factor means that *per* dollar of investment, intermittent energy technologies substitute much fewer MWh of subcritical generation than is the case with HELE coal-fired generation. Therefore, in practice, HELE coal-fired generation mitigates more CO_2 emissions than renewables *per* dollar of investment.[51]

To put this in context, the IEA projects a growth of approximately 10 000 TWh of electricity demand in non-OECD Asia between 2020 and 2040. The analysis in Figure 11 compares the up-front capital investment required for the different generation scenarios which could be used to meet this demand growth.

Investment Option	Generation Mix for 10,000 TWh (%)		Required Capacity (GW)		Total CAPEX[1] ($Billion)	% Increase in CAPEX to Baseline	Annual Emission (Bn. tCO$_2$)
	Coal	Renewable	Coal	Renewable			
Sub-Critical Coal Only	100	0	1,343	0	699	Baseline	9.5
Ultra Super-critical Coal Only	100	0	1,343	0	932	33	7.0
Sub-critical Coal and Onshore Wind	95	5	1,269	241	932	33	9.0
Sub-critical Coal and Solar PV	96	4	1,284	264	932	33	9.1
Onshore Wind Only	0	100	0	4,391	4,944	607	0
Solar PV Only	0	100	0	6,008	6,002	759	0

→ $233 Billion of additional funding required

→ For the same additional financing, ultra super-critical coal technology generates the least amount of emissions

Low load factor renewable technologies means significantly higher required capacity - and therefore higher CAPEX - to generate the same TWh of electricity

Notes:
1) Based on IEA's WEO 2014 New Policy Scenarios capital cost estimates for China in 2035 with construction costs spread equally over the construction period

Source: World Coal Association analysis, 2015

Figure 11 Compared to renewables, High Efficiency Low Emission (HELE) coal technologies can reduce more emissions for the same upfront investment.

In the first instance, this could be met at an investment cost of $699 billion, with subcritical coal-fired generation capacity resulting in 9.5 billion tonnes of CO_2 *per* annum. However, with an extra $233 billion of funding, ultra-supercritical coal-fired capacity could replace all the subcritical capacity, produce the same 10 000 TWh, but emit 2.5 billion tonnes less CO_2 each year.

In contrast, with the same funding, onshore wind or large-scale solar PV cannot displace subcritical coal-fired capacity to the same extent, while also delivering the 10 000 TWh. As a result, the residual energy demand not met by renewable sources may be met by subcritical generation capacity and, as a consequence, onshore wind or large-scale solar photovoltaics (PV) does not reduce emissions by the same amount as ultra-supercritical coal capacity. In other words, no other low-emission generation technology can provide the same high level of generation and low cost as HELE coal-fired power generation.

6.2 Carbon Capture, Utilisation and Storage (CCUS) Development Vital to Meeting Climate Goals

Given society's on-going reliance on fossil fuels, CCUS is vital to achieve the required level of emissions reduction. CCUS will be required not only for coal, but also natural gas and industrial sources to ensure global temperature increases are to be kept below 2 °C.

Yet, the current rate of CCUS deployment is too slow to enable the necessary emissions reductions goals to be achieved. Accordingly, it cannot be expected that developing countries or emerging economies should be the first movers for its deployment. Instead, the international community must concentrate on the following three key areas to support the deployment of CCUS.

6.2.1 International Financing for Carbon Capture, Utilisation and Storage (CCUS). One of the key barriers to the increased deployment of CCUS is the availability of commercial finance. CCUS is currently highly capital-intensive and has associated risks that tend to restrict sources of private finance. While several projects have engaged with debt-finance providers, the Global CCS Institute (GCCSI) suggests that the appetite for, and understanding of, CCS in the financial sector has not been widely tested.

International financing can provide much-needed support in these circumstances. A number of multilateral funding bodies support the development of CCUS projects, including the GCCSI, the World Bank CCS Capacity Building Fund, the Green Climate Fund, the Asian Development Bank CCUS Fund and the UNFCCC Clean Development Mechanism. Existing programmes, however, can only provide some of the capital investment required. More ambitious funding is required by multilateral schemes to improve the financial viability of CCUS projects.

The Low Carbon Technology Partnerships Initiative (LCTPI) launched by the World Business Council for Sustainable Development, together with the Sustainable Development Solutions Network (SDSN) and the IEA offers a particularly promising potential funding model. The platform offers a forum for the diffusion of CCUS by removing barriers and introducing required policy and financial instruments, while promoting Public Private Partnerships (PPPs) on research and development.

6.2.2 Policy Parity. Strong policy drives strong action. Growth in renewable energy technology has been driven by policy that provides $100 billion in subsidies every year.[52] The cumulative value of government policy support provided to CCUS to date is approximately 1% of the cumulative value of policy support provided to renewable technologies. Policy tools available for renewables are not generally made available for CCUS, which has a dampening effect on investment. With strong policy support for CCUS, including parity with other low emissions technologies, the necessary investment will occur. This will make further strides possible in the wider demonstration and deployment of CCUS, which will in turn drive down costs.

6.2.3 Carbon Capture, Utilisation and Storage (CCUS) Deployment Requires International Incentives. Climate solutions require international action. It is imperative that lessons learnt in CCUS projects are shared in international fora, such as the newly developed USA-China Clean Energy Research Centre. In addition, development banks and donors should develop funding mechanisms for CCUS research and development.

7 Coal and Our Energy Future

The global energy system faces many challenges in this century. It will have to continue to supply secure and affordable energy in the face of growing demand. At the same time society expects cleaner energy and fewer emissions, with an increasing emphasis on environmental sustainability.

Alleviating poverty, maintaining secure supplies of energy, and protecting the natural environment are some of the biggest challenges facing our world today. The production and use of coal is linked to each of these challenges.

References

1. United Nations Framework Convention on Climate Change (2016) INDCs as communicated by Parties, Available at: http://www4.unfccc.int/submissions/indc/Submission%20Pages/submissions.aspx.
2. World Coal Association, *Coal and Modern Infrastructure*, World Coal Association, London, 2013.
3. International Energy Agency, World Energy Outlook 2016, p. 550.

4. International Energy Agency, *Technology Roadmap High-Efficiency, Low-Emissions Coal-Fired Power Generation*, OECD/IEA, Paris, 2012.
5. International Energy Agency, Coal Information 2016, p. ii.26.
6. BP, BP Statistical Review of Energy, 2015.
7. Carbon Brief (2016) Mapped: The global coal trade, Available at: https://www.carbonbrief.org/mapped-the-global-coal-trade. Accessed 2 Feb 2017.
8. The Coal Resource: A Comprehensive Overview of Coal, World Coal Institute, London, p. 7.
9. The Coal Resource: A Comprehensive Overview of Coal, World Coal Institute, London, p. 9.
10. International Energy Agency, *Coal Information 2016*, OECD/IEA, Paris, 2016, p. II.8.
11. World Coal Association (2016) Coal mining & the environment, Available at: https://www.worldcoal.org/environmental-protection/coal-mining-environment. Accessed 20 Feb 2016.
12. The Coal Resource: A Comprehensive Overview of Coal, World Coal Institute, London, p. 27.
13. The Coal Resource: A Comprehensive Overview of Coal, World Coal Institute, London, p. 28.
14. The Coal Resource: A Comprehensive Overview of Coal, World Coal Institute, London, p. 10.
15. International Energy Agency, *Coal Information 2016*, p. II.13.
16. International Energy Agency, 2016, *Coal Information 2016*, Paris, OECD/IEA, p. II4.
17. International Energy Agency, World Energy Outlook 2016, p. 207.
18. International Energy Agency, World Energy Outlook 2016, p. 550.
19. International Energy Agency, 2016, *Coal Information 2016*, OECD/IEA, Paris, p. II.15.
20. The Coal Resource: A Comprehensive Overview of Coal, World Coal Institute, London, p. 14.
21. International Energy Agency, *Coal Information 2016*, OECD/IEA, Paris, 2016, p. VI.39.
22. C. Haftendorn, *Economics of the Global Steam Coal Market - Modeling Trade, Competition and Climate Policies*, Technische Universität, Berlin, 2012, p. 19.
23. World Coal Association (2016) Coal & electricity, Available at: https://www.worldcoal.org/coal/uses-coal/coal-electricity. Accessed 20 Feb 2016.
24. The Coal Resource: A Comprehensive Overview of Coal, World Coal Institute, London, p. 20.
25. The Coal Resource: A Comprehensive Overview of Coal, World Coal Institute, London, p. 21.
26. A. Epstein, *The Moral Case for Fossil Fuels*, Penguin, New York, 2014, p. 69.
27. B. Sporton, Coal Helping to Deliver the Four I's of China's G20 Presidency, *G20 Executive Talks*, September 2016, p. 104 [Online]. Available at: http://g20executivetalkseries.com/cover/september-2016/. Accessed 20 Feb 2017.

28. Aurizon, *Sustainability Report*, Aurizon, Brisbane, 2015, p. 16.
29. World Steel Association, *World Steel in Figures*, World Steel Association, London, 2015, p. 17.
30. World Coal Association (2016) How is Steel Produced? Available at: https://www.worldcoal.org/coal/uses-coal/how-steel-produced, Accessed 20 Feb 2017.
31. World Coal Association (2016) Coal & cement, Available at: https://www.worldcoal.org/coal/uses-coal/coal-cement, Accessed 20 Feb 2017.
32. A. Swanson (2015) How China used more cement in 3 years than the U.S. did in the entire 20th Century, Available at: https://www.washingtonpost.com/news/wonk/wp/2015/03/24/how-china-used-more-cement-in-3-years-than-the-u-s-did-in-the-entire-20th-century/?utm_term=.8c7bba0e6a4d. Accessed 20 Feb 2017.
33. World Coal Institute, Coal Liquid Fuels, World Coal Institute, London, 2011.
34. International Energy Agency, World Energy Outlook 2016, p. 92.
35. International Energy Agency, World Energy Outlook 2016, p. 93.
36. R. Bryce, Not Beyond Coal. How the Global Thirst for Low Cost Electricity continues driving coal demand. Centre for Energy Policy and the Environment at the Manhattan Institute. N° 14 October 2014. Executive Summary.
37. International Energy Agency, *World Energy Outlook*, OECD/IEA, Paris, 2016, p. 552.
38. International Energy Agency, *World Energy Outlook: Energy for All*, IEA/OECD, Paris, 2011, p. 26.
39. U. Bhaskar, India's per capita electricity consumption touches 1010 kWh, *Livemint*, 20 July 2015.
40. World Coal Association, *India's Energy Trilemma*, World Coal Association, London, 2014.
41. BP, *BP Statistical Review of Energy*, BP, London, 2016.
42. G. Sheridan (2016) India's challenge is 24/7 electricity for all, Available at: http://www.theaustralian.com.au/opinion/columnists/greg-sheridan/indias-challenge-is-247-electricity-for-all/news-story/178dc1ee3efdfecd61e23afc335f5d1f. Accessed 20 Feb 2017.
43. World Coal Association, *The Power of High Efficiency Coal*, World Coal Association, London, 2016.
44. Coal Industry Advisory Board, *The Socio-economic Impacts of Advanced Technology Coal-Fuelled Power Stations*, CIAB, Paris, 2015.
45. World Bank Group, Enterprise Surveys, India, 2014.
46. Coal Industry Advisory Board, *The Socio-economic Impacts of Advanced Technology Coal-Fuelled Power Stations*, CIAB, Paris, 2015, p. 37.
47. Coal Industry Advisory Board, *The Socio-economic Impacts of Advanced Technology Coal-Fuelled Power Stations*, CIAB, Paris, 2015, p. 40.
48. Chamber of Mines of South Africa, Annual Report 2013/2014, May 2015, p. 3.

49. Africa Development Bank Group, ESIA summary - Nacala rail and port project.
50. World Coal Association, *Platform for Accelerating Coal Efficiency*, World Coal Association, London, 2014.
51. World Coal Association, *The Power of High Efficiency Coal*, World Coal Association, London, 2016.
52. World Coal Association, *Carbon Capture and Storage – The Vital Role of CCS in an Effective COP21 Agreement*, World Coal Association, London, 2015.

Coal Mining

ZESHAN HYDER

ABSTRACT

Coal mining is an important industry that contributes to the economic development of a country. There are two main coal mining methods: surface mining and underground mining. In the surface mining method, coal is first exposed to the surface and then mined with heavy equipment, whereas in underground mining, coal is mined *in situ* by accessing through shafts, slopes and tunnels, and mined coal is brought to the surface. Coal mining provides several economic benefits including employment, tax revenue, improved infrastructures, development of support industry and GDP growth. However, coal mining, transportation and consumption have serious environmental consequences ranging from nuisances in the form of dust, noise and increased traffic to greenhouse gas emissions, acidic mine drainage and subsidence.

1 Introduction

Coal mining, as an industry, is a major contributor in the development of a country. Coal mining played a significant role in the development of current industrial heavyweights like Britain, Germany, Australia, the United States and Europe.[1] Despite a decline in demand owing to environmental legislation and regulatory restrictions, coal remains one of the largest sources of energy world-wide. In 2016, coal was the second-largest energy source worldwide after petroleum and other liquids. This trend is expected to continue until 2030.[2] The largest producers of coal are China, the

Issues in Environmental Science and Technology No. 45
Coal in the 21st Century: Energy Needs, Chemicals and Environmental Controls
Edited by R.E. Hester and R.M. Harrison
© The Royal Society of Chemistry 2018
Published by the Royal Society of Chemistry, www.rsc.org

United States, India, Australia and Indonesia.[3] Traditionally, most of the coal produced by a country was consumed locally and a mere 15% of coal was exported to other nations; however, since 2000 this trend has changed and total coal exports have more than doubled since 2000.[4,5]

1.1 Brief History of Coal Mining

No recorded evidence exists of when coal mining first started; however, it is presumed that the discovery of a black stone burning when exposed to fire may have occurred purely accidently in different parts of the world over thousands of years. The first recorded use of coal in China occurred 1100 years before the Christian era.[6] Coal use can be traced back to prehistoric times; however, the dominance of coal as a fuel dawned in the 18th century. Britain was a dominant user of coal and by 1760 there were about 17 coke-fired blast furnaces in Britain.[7] Coal smelting then moved to other nations and this process was recorded to occur in France in 1785, Silesia in 1791, Belgium in 1823, Austria in 1828 and the Ruhr in 1850.[7] Coal use as a furnace fuel can be dated back to the Han dynasty in China.[8] The first known record of coal in Nova Scotia dates back to 1686 when Jean-Baptiste–Louis Franquelin first observed coal there.[9]

In the USA, Hopi Indians used coal to bake their clay pottery in 1000 AD. In 1748, the first recorded commercial mining of coal occurred in the USA in the Manakina Area.[10] As makes sense, the earliest method of extracting coal was surface mining, where coal was mined from outcrops and deposits exposed to the surface.[6]

2 Coal Mining Methods

There are two major methods of extracting coal from the ground: surface mining and underground mining. The selection of surface or underground mining method depends upon several factors, including: depth; thickness; inclination or dip of coal seam and surrounding strata; quality of the coal; geology and topography of the area; economic impact and cost of selected method; geomechanical conditions; presence of water bodies and aquifer; distance from local population; availability of infrastructure and workforce; environmental impacts; and reclamation, restoration and closure requirements.[11]

Another important technical factor in the selection of the mining method is the *stripping ratio*.[12] Stripping ratio is defined as the amount of overburden that has to be removed to produce one production unit of coal. In the USA, the most common unit of production is the ton and the amount of overburden removed is usually measured in cubic yards (cubic meters) or tons, so stripping ratio is expressed either in cubic yards (cubic meters) *per* ton or tons of overburden *per* ton of coal. Generally, a higher stripping ratio favours underground mining whereas lower stripping ratios favours surface mining.

2.1   Surface Mining

In surface mining, the coal deposit to be extracted is generally exposed to the surface. Surface mining is a widely used method for exploitation of minerals worldwide;[13] however, for coal, underground mining is the dominant method of exploitation.[14] Still, in some major coal-producing countries surface mining is more common than underground mining.[15] In the USA about 85% of all minerals are mined using surface mining methods.[16] About 98% of metallic ores, 97% of non-metallic ores and 61% of coal in the USA are mined using surface mining methods. In 2015, the USA produced approximately 897 million tons of coal, of which about 66% was produced using surface mining methods and 34% through underground mining.[16] Figure 1 shows the productive capacity of US coal mines by mine type.

2.1.1   Surface Mining Methods. Surface mining methods are typically applied when the coal deposit is close to a surface that is flat or has only a gentle slope. Topsoil and rock-covering removal is typically the first step in surface mining. Coal is first exposed to the surface by removing vegetation, overburden soil, rocks and any other material covering the deposit. Modern surface mining methods are highly mechanized and they apply some of the largest machines available in any industry.

Three types of open cast or surface-mining methods are common for exploitation of coal deposits: *open-pit mining*, *strip mining* and *mountain-top removal*. Of these, strip mining is the most widely used surface mining method.[17]

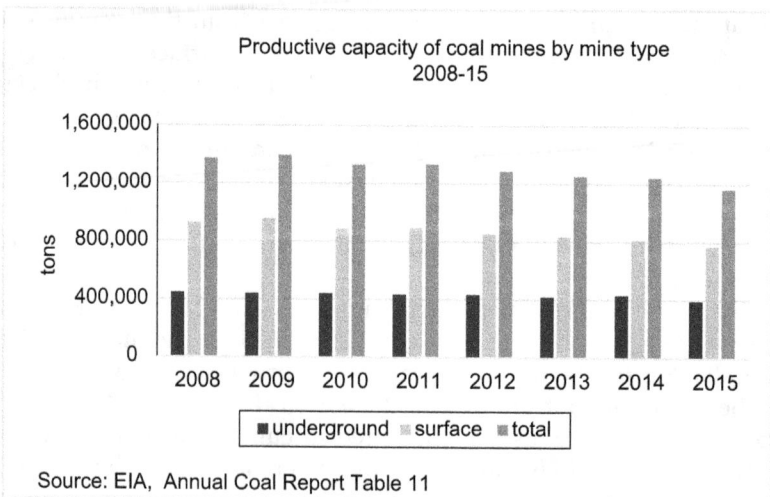

Source: EIA, Annual Coal Report Table 11

Figure 1   Productive capacity of US coal mines by mine type, 2008–15.
(Source: US Energy Information Administration, November 2016).

2.1.2 Open-pit Mining. Open-pit mining is used when the coal deposit is either outcropping, *i.e.* exposed to the surface, or is at a shallower depth, typically less than 200 feet.[18] If coal is buried deeper than that, then open-pit mining is not considered a feasible method and underground mining methods are applied to exploit that deposit.

In open-pit mining, coal is mined using a series of benches. These benches are developed top down and take the form of a series of horizontal layers, generally of uniform thickness,[19] as shown in Figure 2.

Mining starts from the highest bench and continues sequentially to the lowest bench that can be developed safely and economically. This lowest bench is termed as the *final pit limit* or *pit bottom*. This final pit limit depends upon economic and safety considerations. Some of the factors that play an important role in limiting the final pit outline are the stripping ratio; slope stability; structural geology of the area; strength of surrounding rocks and strata; type and capability of machinery being used; presence of water or aquifer; access road grade; and overall pit slope angle.[20–22] After the final pit limit is reached, the open pit looks like an inverted spiral.

Access to these benches is generally through a road which is most commonly termed an *access ramp*. The width and steepness of this ramp depends upon the type, size and number of machines and other equipment used in the mine.[23] This ramp goes winding and spiralling through different benches and provides access for transportation of coal, waste material and equipment.

An open-pit mine is a highly mechanized complex structure, which starts with the bench as a basic extraction component. Each bench has a typical geometry that is characterized by bench height, width and slope. Bench height is the vertical distance between the topmost point of the bench, termed the *crest* and the lowest point of the bench, termed the *toe*. A typical bench height varies between 8 and 25 meters.[24,25] Bench height depends upon type of overburden rock: structural geology of the strata, *i.e.* number of

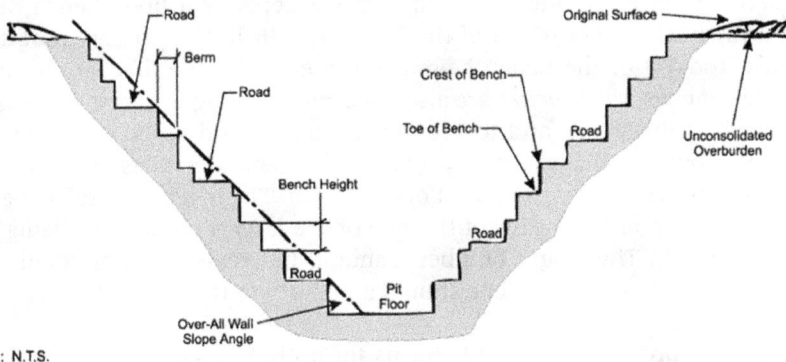

Figure 2 Open pit mine design (NTS = not to scale).
(Open access source: Vale Inco, available at http://www.vbnc.com/eis/chap3/chap3.htm#Continue).

joints; faults and fissures present; presence of water; type and size of equipment being used; extraction method applied, *i.e.* conventional drilling and blasting or mechanical excavation; and bench slope angle and slope stability considerations. The total number of benches in a mine depends upon the depth and thickness of the deposit. A thin and shallow deposit requires only a few benches whereas for thick, deeper deposits many more benches are required.

The bench face is the exposed vertical or sub-vertical slice between the crest and toe of the bench. Bench slope angle is one of the most important design parameters of an open-pit mine and depends upon stability and safety requirements. Bench slope angle is defined as the average angle that a bench face makes with the horizontal and ranges between 55° and 80°, depending upon the hardness, strength and geological characteristics of the rock.[19] The steeper the bench is, the lesser is the overburden to be removed. However, steepness of face angle is governed by safety and slope stability requirements.

The lower exposed surface of the bench is termed the *bench floor* and is generally the working platform. The horizontal distance between the crest of a bench and the toe of that same bench is termed the *bench width*. The bench width typically depends upon number; size; width and type of excavation; and the loading and haulage equipment used.[26] Depending upon nature of operations, benches are divided into working benches, safety benches and *berms* (see below for explanation). A working bench is one at which mining activity is in progress. Depending upon the design of the open pit, there may be more than one working bench at the same time, or different operations may occur at different locations of the same bench. For example, it is typical for the same working bench to be in the process of development, drilling, blasting, excavating, loading and hauling at the same time, at different locations.

Safety benches are left on every level during primary extraction to collect the material sliding from the upper benches and to stop the downward progress of large boulders.[19] Their width is dependent upon bench height, typically in the order of 2/3 of the bench height in the primary production phase and 1/3 of the bench height at the end of mine life. In addition to safety benches, *safety berms* are also constructed along the crest of the bench to catch falling rocks and to prevent the trucks and other machines from backing over.[27,28] These safety berms are generally piles of waste rock material collected along the crest of the bench. The height of a safety berm is dependent upon the radius of the tyre of the largest equipment being used on the bench. The height of a berm should be greater than or equal to the radius of the tyre and its slope should about 35° or the angle of repose of the waste material.[23,28]

Benches provide working platforms for both men and machinery for the extraction of coal. The development of benches starts with the removal of surface vegetation, trees, grass, topsoil and other overlying structures covering the upper layer of the designed open-pit mine. Topsoil is removed and stored to be later used in the reclamation and rehabilitation phase. Generally bulldozers, scrapers, shovels and scrapers are used in the process of topsoil removal.[29] After topsoil is removed and stored, the subsoil is removed from the designed area of the open-pit mine. This subsoil is more commonly termed as *overburden* and waste material. This is the overlying rock or material that has to be removed to gain access to the coal deposit. Typically, the first few benches are driven into the overburden. This overburden is extracted and hauled out of the mine area in haul trucks and dumped outside the final pit boundary in the form of waste spoils.

The final pit limit is based on the *overall pit slope angle* and maximum allowable stripping ratio. The overall pit slope angle is the angle measured from the crest of the topmost bench to the toe of the pit bottom. This angle depends upon the number of benches, bench height, bench width, individual bench angle and width of the ramp or haul road. As the allowable pit bottom is reached based on stripping ratio, overall pit slope angle and material handling limitations, a careful economic and technical analysis is carried out to decide whether to continue mining using underground methods or stop the mining activity and start mine closure and the reclamation phase. The world's largest surface coal reserve is at Peabody Energy's North Antelope Rochelle Coal Mine located in Wyoming in the USA.[30]

If it is decided to suspend mining activity, then the reclamation and restoration phase starts. In this phase, the overall pit boundaries are secured using permanent fencing, the topsoil is re-spread and compacted using heavy machinery and, if possible, subsoil is hauled back into the pit. After placement of topsoil, the re-vegetation process starts and attempts are made to restore the original landscape, topography and contours of the area as much as technically possible.[31]

2.1.3 Strip Mining. Strip mining is the most common surface mining method used for exploitation of coal in the USA. Strip mining is feasible for flat-lying, bedded deposits that are close to the surface. Bedded deposits are accumulations of coal in layers or flat strata that are more or less parallel to each other. Coal deposits that are shallow and have gentle slopes of less than 15° are mined using strip-mining methods.[11] As the name implies, the deposit is worked in the form of long strips. The strip is generally much larger in length, sometimes reaching about 1000 m, whereas the width of a single strip ranges between 60 and 120 m, depending upon the dimensions of the excavation equipment.[32]

Strip mining is a highly mechanized method and uses some of the largest machines on earth.[33,34] Stripping starts with site preparation and the removal of surface vegetation. After removing the vegetation, the topsoil is scraped away and removed for storage, either within the mining area or some close-by location to utilize at the end of mining. After topsoil removal, overburden is removed using very heavy excavation equipment. If the overburden is hard and difficult to excavate, drilling and blasting is applied to fragment and loosen the rocks and overlying material.[35]

The removal of overburden starts with the creation of a deep cut in the top surface. This cut is generally termed a *box cut*. This box cut is further expanded with excavation equipment and forms the top-most bench of the mine.[36] The overburden material thus removed is placed within the area of the mine in the form of a waste dump of designed slope and width. This material is later on casted back in the strip area from where coal has been mined. After the first bench, ranging in height from 30 to 40 m, advances a considerable distance, a second bench in the same strip is propagated, either in the overlying overburden if still there is some, or in the coal seam. At the same time several benches are worked together, some excavating overburden material while others are removing coal. These benches are developed up to the designed pit limit.

The coal removed this way is taken away either in haulage trucks or on conveyor belts.[37] Conveyor belts are the most common means of coal haulage in strip mining. The overburden or waste rock is stacked alongside the strip currently being worked in the form of waste spoils. After the bottom-most strip of coal has progressed a considerable distance, the reclamation process starts concurrently. The waste spoil is casted back into the excavated area and the backfilling operation starts. After backfilling the waste spoil into the excavated area, the soil is re-surfaced, re-contoured, re-graded and compacted. The topsoil that was stored during the initial excavation is hauled back and spread over the entire re-surfaced area. After topsoil placement, the final re-grading and compaction operation begins. After this stage, vegetation activity starts and the area is restored to match more or less the original landscape and transformed into a useful area for recreational, commercial or agricultural use.[38]

Strip mining utilizes some of the largest and most modern mining machines. Some detail and uses of these machines is as follows:

2.1.3.1 *Excavation Equipment*

Excavation machines are used for removal of either overburden rock or coal itself. These are high capacity, very rigid, usually of very limited mobility and very high-production machines. They are used to cut, dig and load the excavated material.

Shovels, sometimes also termed *backhoes*, based on their design, are the most common surface-mining equipment and are used for a variety of applications. These are bucket-loaded machines with very high bucket capacities. There are two main types of shovels in terms of powering them: electric shovels and hydraulic shovels. Electric shovels have bucket capacities ranging from 15 to 100 m^3 and are operated by electric motors. They are very rigid, less mobile and high-production machines. An electric shovel with a bucket capacity of 100 m^3 can fill a 300-ton haul truck in about three passes.[39] Hydraulic shovels compared to electric shovels are more mobile and agile but have less capacity.[40] Most shovels are crawler mounted. Crawler mounted shovels do not use rubber tyres; instead, they are equipped with their own tracks and move on those tracks. They are used for cutting, digging, loading, trenching and landscaping purposes.

Draglines are essential equipment for surface mining operations in general and for stripping in particular. Their ability to dig above and below their level makes them ideal for strip mining. They have the ability to cut through overlying, as well as underlying, rock and strata. These are high-capacity rigid machines used for handling unconsolidated and softer materials.[41] If rock is hard, then it is fragmented using drill and blast before dragline operation.

Scrapers are the most mobile of the surface mining equipment, but their use is limited to soft and well-fragmented material. They are used in scraping, stripping, levelling and landscaping purposes.

Bucket wheel excavators are the largest of surface mining equipment, with very high-production capacities. They are capital-intensive machines and are generally limited to soft and easy-digging ores and strata and are ideal for coal mining. They are continuous digging machines that contain a large wheel mounted with a series of buckets that dig the earth and scoop the material in a continuous manner. They are used for removing overburden and excavating exposed coal deposit.

Surface drills are used as a supporting equipment in combination with other excavation equipment. They are used to drill holes in the hard rock and strata. These holes are then loaded with explosives and blasted to fragment the rock. This fragmented material is then easier to handle by excavation and haulage equipment.

2.1.3.2 Haulage Equipment

Haulage equipment is used in transporting waste and mined material. Waste material is either dumped close to the mined area for recasting or is transported to the spoil tip. Mined coal is transported to either a processing/coal-washing facility or to the coal stockpile. Haulage equipment is an essential part of the excavation equipment as it transports the excavated material and provides space for continuous operation. Some of the most common haulage equipment utilized in surface mining operations in general and strip mining in particular is as follows:

Bull dozers are crawler or tyre-mounted machines equipped with a blade used to scrap, excavate and push large quantities of soil, sand, dust and loose and fragmented rocks. Some bulldozers are also equipped with a ripper that is used to loosen compacted soil. They are mostly used as an auxiliary equipment in strip mining, helping in the removal of overburden, ripping, scrapping, recasting, levelling, re-surfacing, re-grading and compaction of loose soil for re-vegetation.

Haul trucks are the main hauling equipment in shovel excavation. These are very mobile, high-capacity and rigid machines that can travel up and down steep ramps. They require well-maintained access roads for their operations and are generally worked in fleets. They can transport the mined material over fairly large distances but their economic cost increases considerably with increasing haul distances.[42]

Trains require specially built tracks to access the material; however, they provide very high-volume transportation over long distances and have relatively low operating costs. Against this, their initial capital cost is very high but they are ideal for transporting final product over long distances. Their biggest limitation is steep grades. Trains cannot handle a grade steeper than 3°.[43,44]

Conveyors provide a high-volume continuous hauling method that has high initial capital cost but low operating costs. They provide continuous transportation of waste and coal over considerably long distances and can tolerate variations in slopes. They are an essential part of a bucket-wheel excavator system. Their mobility is very low and they require that the material to be hauled is fairly fragmented. Modern in-pit conveyers are finding wide application in open-pit surface mines as a hauling alternative over some segments of the pit.

2.1.4 Special Mining Methods (Surface). *Mountain top removal mining* is a surface mining method where coal that is either exposed at the top of a mountain in the form of an outcrop or buried closer to the ridge of the mountain is exploited by exposing it to the surface. Mountain top removal is a controversial mining method as it involves deforestation of the desig-nated mining area. The vegetation and tress in the area are scrapped and land is cleared for mining activity. The overlying rock is removed using drilling and blasting and heavy excavation machinery.[45] Coal is exposed by removing overburden and then mined using draglines, shovels and other heavy equipment. Waste material is stored in nearby fill locations and later on is used during the reclamation phase. The reclamation activity requires backfilling the waste material into the mined area and restoring the contour to the maximum extent technically practicable. Figure 3 illus-trates the mountain top removal mining method.

Contour mining is a special type of mountain removal mining where coal outcropping at different elevations across the mountain is removed by cre-ating one or more benches close to the outcrop. At the same time, several benches can be worked across the same mountain.[46] After access roads are built, a flat area or bench is developed to provide a platform for the heavy equipment to operate.[47] The overburden is removed and stored locally and coal is mined. After mining activity, the overburden or waste is filled back and contour is restored. In contour mining, the overburden removal and deforestation is localized and limited to the outcropping seam only.

Auger mining is a method used mostly to supplement contour or strip mining to remove the coal that cannot be removed by these other methods because of excessive overburden removal requirements. Auger mining in-volves drilling of large diameter horizontal holes into the exposed coal seam and recovering as much coal as technically and safely possible.[48,49] Auger mining occurs directly at the bottom of exposed high walls; historically, of

Figure 3 Mountain top removal mining.
(Reproduced, with permission, from http://ohvec.org/high-resolution-mountaintop-removal-pictures/; Photo: Vivian Stockman/ohvec.org. Flyover courtesy SouthWings.org).

greater concern has been the safety of workers rather than environmental impacts.[50]

2.1.5 Economic Impacts of Surface Coal Mining. Mining is an important industry that impacts both the regional and national economy. Mining, like any other industry, has both positive and negative economic and socio-economic impacts. Some pronounced economic impacts of mining are as follows:

Increased employment opportunities, especially for local and regional populations, are one of the biggest economic impacts of mining. These provide better opportunities for employment, better living standards and higher incomes for the local communities.[51] Mining provides comparatively high-paying jobs. Mining not only provides direct employment but also a source of raw material for many other industries, resulting in a growth of the industrial sector in the region and providing many indirect jobs. In addition to industrial development opportunities, mining causes a surge in population, growth of service industry, local stores, businesses, retailers and hotels, *etc.*[52,53] Direct employment by the surface coal mining sector in the USA in 2015 was approximately 25 000 employees.[54] One criticism of the economic impact of mining is that it provides only a short-term economic boom.[55] After major coal activity declines or the life of a mine ends, the population decreases drastically, major migration occurs and the area suffers as it not only loses its major economic stimulus but traditional economic activity generated during that boom also declines. In addition to loss of natural resources, traditional jobs and local industry, environmental degradation continues for years after a mine is abandoned.[56]

Coal also provides a source of taxation income and revenue for local, state and national governments. Coal is a major export in several developing countries where it provides valuable foreign exchange that can be utilized in both local and national development.[52]

Coal is one of the largest energy sources worldwide, primarily in the production of electricity. The electric power sector was the largest consumer of coal in the USA, using 852 million tons in 2014 and 738 million tons in 2015. Coal is still the largest fuel consumed for electricity generation worldwide.[57]

2.1.6 Environmental Impacts of Surface Coal Mining.   One of the biggest criticism of surface mining is its large environmental footprint.[58] Mining, by its design, is damaging to the environment as it consumes non-renewable natural resources; produces greenhouse gases, organic and inorganic materials; releases waste material; contaminates water; and destroys surface vegetation and forests.[53,59] Surface mining has a much larger footprint than underground mining.

Surface mining starts with the removal of surface vegetation, plants, forests, and any installations and structures within the mining area. The waste material and overlying strata are fragmented using explosives and heavy machinery that release harmful gases into the atmosphere. Coal deposit is exposed to the open atmosphere and can release harmful gases, toxic minerals, heavy metals and harmful materials in the atmosphere and in the ground.[60] Migration of toxic and harmful mineral matters, sulfur, mercury and radioactive material to the surface and sub-surface water resources is a threat to the health of the populace and resources of potable water.[61]

Waste that can release harmful materials is stored on the surface. One major problem arising from the surface exposure and storage of coal is acidic mineral-rich water draining to the nearby land, soil and water bodies.[62] Acidic mine drainage is one of the prominent environmental problems caused by mining activities and poses a big challenge in establishing environmentally responsible mining. Over the last few decades several attempts have been made to mitigate this problem.[63–69]

Surface mining causes changes in landform by changing topography, creating new hills and mountains of waste and tailings, resulting in landslides and slope failures; causing subsidence and earthquakes; and creating a visually unpleasant site.[58]

Surface mining causes deforestation, as, for operations to commence, the land has to be cleared of vegetation, grassland, plants, shrubs and trees. This not only results in the degradation of land, limiting its future uses, but it is also harmful to biodiversity. An integral part of any post-mining reclamation and restoration plan is re-vegetation that includes mulching, irrigation, fertilization, direct topsoil hauling in the mined area and promotion of landscape diversity.[70] Surface mining activities result in increased erosion of land and increased soil discharge into rivers and streams, causing changes in the dynamics of stream channels; increased turbidity; loss of biodiversity; and alteration of food-web structure.[71,72]

2.2 Underground Mining

Underground mining consists of a number of methods for exploiting coal and other minerals that are located below the surface of earth. These mining methods are applied to coal deposits that cannot be mined through surface mining methods and are located at depths of more than 200 m, or are at shallow depths but which are not technically feasible to be mined through surface methods. Underground coal mining has a relatively smaller production and productivity than surface coal mining but it also has a smaller environmental footprint.

The deposit suitable for underground mining is accessed through vertical openings known as *shafts* or horizontal openings termed as *slopes*, *drifts* and *tunnels*.[73] These provide access for men, machinery, electric and power cables, mine hoists and haulage equipment, and ventilation air supply. Shafts are generally equipped with an elevator, termed a *cage* or *skip*, which moves men and material vertically. This vertical movement between underground mine and surface facilities is termed *hoisting*.[74]

2.2.1 Underground Mining Methods. The most common underground mining methods are *longwall mining* and *room and pillar mining*. Room and pillar was very common in the USA, but longwall mining is of growing importance in the mining industry.[75] Coal deposits suitable for underground mining are deep-seated, relatively flat-lying and generally tabular, *i.e.* table-like or stratified in extent.

Underground mining is not as extensively mechanized as surface mining and the size of machines is much smaller. Underground mines are of three types: *non-mechanized* or *manual*, *semi-mechanized*, and *fully automated and mechanized*. Most of the underground mines in developing countries are semi-mechanized or manual, whereas most developed countries like the USA, Australia, UK and other European countries have highly mechanized underground coal mines. The advancement and sophistication of mine machinery and production equipment is giving rise to increasing mechanization and automation in underground coal mines.

Underground coal mining, like surface mining, faces several technical and engineering challenges. These include: ground control and support requirements; strata gases and water control; provision of sufficient quantities of air for ventilation and pollutant dilution; methane drainage problems; rock and coal bursts; heat and thermodynamic control; and operations at increasing depths.[76–79]

2.2.2 Longwall Mining. Historically, longwall mining has been a popular method of mining, developed mainly in the UK and Europe. Although longwall mining was introduced in the USA in late nineteenth century, it did not gain much popularity until later.[80] The development of mechanization in longwall mining helped its growth in the USA. In the last few decades, longwall mining has developed as a high-production, highly-mechanized coal mining method in the USA; in 2005, there were only 53 longwall mining faces but they accounted for 52% of underground coal production.[80]

Longwall mining starts with the development of an access shaft, slope, or drift to reach the deposit. From the main shaft, secondary openings are driven that connect the main shaft to the coal deposit or panel. These secondary entries are termed *crosscut*. After these openings are driven, a long panel of coal is prepared for mining. Across the panel are a series of long entries, termed *main entries*, which are used for transportation of coal, machinery, equipment and ventilation air. These main entries divide the coal block into several panels that have a width of 200 to 400 m and a length of 3 to 5 km.[80] Due to the form of these large panels, this method of mining is termed longwall mining.

After the development of panels, large shearers start mining the coal. The mined-out coal is loaded onto conveyor belts running across the working face. These conveyors take the coal either to the shaft bottom for hoisting to the surface *via* cages and skips, or transport it to another conveyor running through a slope to the surface. In some cases, haul trucks are also employed to carry waste and coal to the surface through transportation drifts or slopes.[81]

Longwall mining is very high-production and productivity method. However, this method is highly capital intensive and development activities have a long lead-time before actual coal production. The development is sometimes carried out more than two years in advance of production and, therefore, makes it a rigid method as changing the layout after such development is very difficult.[80] This method requires a relatively flat, continuous, tabular block of coal with large dimensions and fewer fluctuations in strata and geology.[6]

Based on the direction of mining, this method is divided into two types: *longwall advancing* and *longwall retreating*. Longwall retreating is more common than longwall advancing; however, it requires the injection of large amounts of capital before the start of actual coal production.[6,82] In this method, entries and panels are developed to the designed extent of the mine and the coalface advances in a direction opposite to the initial panel development, hence the term longwall retreating. Mechanical shields are used to provide support for the working face, shearer and overlying strata. After the coalface has advanced a designed distance, the overlying strata are allowed to collapse, either by removing the mechanical support or by using explosives to aid controlled collapse. This method is also known as the *caving method* because the strata in the mined-out area are made to cave in to the space created by the mining activity. Figure 4 illustrates the longwall retreating method.

In longwall advancing, the longwall face follows the developmental entries and is sequentially driven behind the excavation of entries and panels. The

Retreating Panel Layout

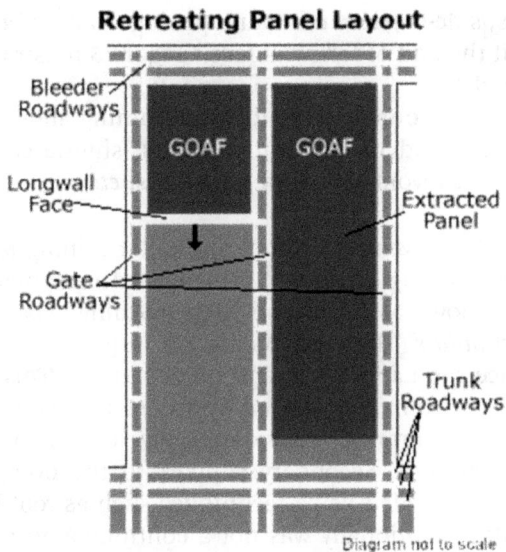

Figure 4   Longwall retreating (GOAF is the area from which coal has been removed). (Open access source: University of Wollongong Australia, available at http://eis.uow.edu.au/longwall/html/retreat.html).

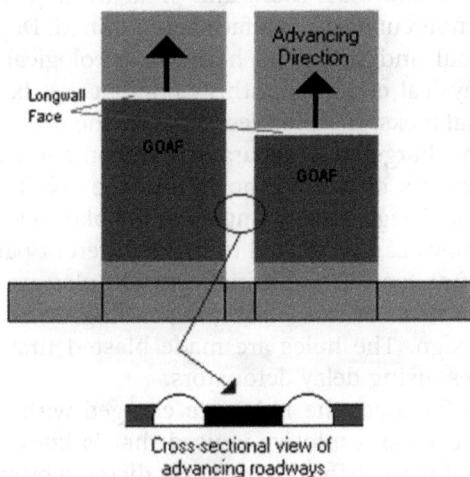

Figure 5   Longwall advancing (GOAF is the area from which coal has been removed). (Open access source: University of Wollongong Australia, available at http://eis.uow.edu.au/longwall/html/advance.html).

longwall face and developmental entries move in the same direction, which is why this method is termed the longwall advancing method. The advantage of this method is a smaller lead period before actual mining and cash flow starts, and that reduces the initial capital requirement.[82] Disadvantages are that the galleries, side-entries and main openings have to be supported and maintained until the end of longwall face.[6] Figure 5 illustrates the longwall advancing method.

The machines commonly used in longwall mining are shearers, continuous miners, shields and mechanical supports, shuttle cars, loaders, belt conveyors, chain conveyors and slope conveyors, *etc.*

2.2.3   Room and Pillar Mining.   Room and pillar mining is most common underground coal mining method.[83] It involves development of a series of mined-out areas known as *rooms* and large un-mined coal blocks between the rooms called *pillars*. These pillars are left in place to support the overlying strata; hence the name room and pillar mining. This method is also termed the *unsupported mining method* as it does not require large primary wooden or steel supports. The systematically left-over un-mined pillars provide primary support. Secondary support to the overlying strata and mined openings is provided using techniques such as roof bolting.

Room and pillar traditionally was not a continuous form of mining; rather, it was a cyclic mining technique. However, with the advent of modern coal-cutting machines and other advancements in mine machinery, room and pillar can now be classified into two broad categories: *conventional drill and blast mining* and *modern continuous mining*.

Conventional drill and blast room and pillar mining starts with the development of an undercut at the coalface being mined. Depending upon the strength of the coal and strata, its hardness, geological and mechanical properties and physical conditions, this undercut can be developed using explosive or manual picks. If explosives are used, the undercut may be a part of the production charge or a separate developmental charge. The production round consists of drilling holes into the coalface according to a predetermined blast design. Depending upon the blast design requirements, the central blast holes can be drilled to make different patterns in order to create additional free face. These initial central holes are termed as *V-Cut*, *pyramid cuts*, or *burn cut* or *snake holes*, depending upon their orientation, alignment and design. The holes are made blasted first using sequential blasting techniques, using delay detonators.

After drilling is finished, the holes are charged with explosive. In coal mining, a special class of explosive is used that is known as a *permissible explosive*. The use of permissible explosives is dictated by safety legalisations to avoid unintended fires, explosions, production of harmful gases and other hazardous conditions. After loading, the explosive is blasted using either electric or non-electric accessories.

After blasting, the fragmented material is loaded into mine cars and shuttles and is transported to the surface. In the meantime, the roof and sides or ribs are supported if needed by rock bolts, wooden or steel supports. After the fragmented material is hauled away, the next round of drilling holes starts and work continues in this cyclic manner.

To give continuity to the cyclic nature of the conventional drill and blast method, several faces are worked at the same time. In some, the drilling operation is carried out, while in others explosive loading, fragmented-coal haulage and roof-supporting operations are in process.

The development of continuous miners and coal-cutting machines has revolutionized underground coal mining and cyclic room and pillar mining has been transformed to a continuous operation. Continuous miners break coal from the face. The coal is transferred to shuttle cars *via* an in-built conveyor system and then hauled to the surface. The roof-bolting machine is an important part of the system and provides support for the mined-out area. Continuous miners can excavate the coalface and produce coal at a rate of 5 tons min^{-1}. The term *continuous mining* is strictly a misnomer. Although a continuous miner can operate continuously, it has to wait for the roof support to be completed and sometimes wait for the shuttle cars to return for re-loading. Thus there is a significant delay in cutting a pass, loading and hauling, roof support and further cutting. Efforts are being made to reduce these delays and make the process more productive and less time-consuming.

After rooms have been developed and pillars have been left in place, at the end of a mine's life more than 50% of coal still remains in place in the form of pillars. The final mining operation, therefore, is extraction of these pillars. This process is termed as retreat mining.[84] Although the pillar extraction can technically be carried out simultaneously with the production mining, it is most common to start retreat mining at the end of production mining. Retreat mining or pillar extraction is a hazardous operation and requires rigorous technical skills and extensive planning.[85–87] It is conducted under strict regulatory and safety controls.[88]

Even after pillar extraction, a large percentage of the coal still is not re-coverable and remains underground, tied up as barrier pillars, support pillars and pillar that are difficult to extract due to technical and safety concerns. Rock and coal burst, stress redistributions, stress loading and catastrophic failure of pillars are some of the hazards associated with pillar extraction.[89] For coal, the usual recovery rate is about 60%, *i.e.* about 40% of the coal remains in place and cannot be excavated using current mining techniques.[90] This is a big loss of natural resource as this coal is almost impossible to recover and use in the future, even with advancement in mining technology.

2.2.4 Hybrid Mining Methods (Underground). Hybrid mining methods are a compromise between fully automated continuous mining and cyclic conventional mining methods, or between surface and underground mining methods. In some instances, mining starts as a surface mining operation but, based on technical, safety and economic considerations, it becomes converted into an underground mine and the remaining deposit is recovered though shafts, drifts or slopes.

2.2.4.1 Shortwall Mining
Shortwall mining is a hybrid of room and pillar and longwall mining methods.[91] Initial development is similar to room and pillar, using the same

equipment employed in this type of mining; however, the pillars are much larger in size than in conventional room and pillar mining.[82] These pillars are extracted in panels resembling longwall panels, but these are much smaller in size. The overlying roof is controlled using heavy automated supports as applied in longwall and the area behind the supports is allowed to collapse. A typical shortwall panel is about 250 ft wide.[82]

2.2.5 Economic Impacts of Underground Coal Mining. The economic impacts of underground coal mining are similar to surface mining. These include: increased employment opportunities; increased source of income for local communities; development of support industries; increased commercial activities for the local populace; increased incomes for local, state and federal governments resulting from increased tax revenues and royalties; improved purchasing power; low cost energy; and GDP growth. Negative economic impacts of underground coal mining include: lowering of property values; decreased land availability for agricultural use; reduction in local indigenous industry and businesses because of mining industry growth and better earning opportunities; deterioration of roads and increased maintenance requirement of local roads and infrastructure; population increase; and increased traffic and congestion.

2.2.6 Environmental Impacts of Underground Coal Mining. The environmental impacts of underground mining are less than those of surface mining as underground mining has much smaller surface requirements and most of the activity is underground without much exposure to the surface during the production phase. Nevertheless, underground mining has its own environmental impacts, which include: increased dust, noise and air pollution; hazardous waste; surface and underground water pollution; acidic mine drainage; surface and underground coal fires; increased greenhouse gas emission (GHG) emissions; subsidence; surface deterioration; surface waste piles; coal storage and tailings disposal.[92]

2.3 Novel Methods

These are not typical mining methods but are innovative ways to utilize coal and related products without mining coal. In these methods the typical mining activity of removing coal through a series of surface and sub-surface openings, shafts, slopes and tunnels is absent.

2.3.1 Underground Coal Gasification. Underground coal gasification (UCG), also known as *in situ* coal gasification, is a unique underground coal mining method. In this method, instead of gaining access to underground coal deposits through shafts, drifts, tunnels and other in-ground entries, coal is converted into a gaseous product *in situ* by injecting oxygen or air into the deposit through a series of injection wells. This converts the coal into a gaseous product commonly known as *syngas* or *synthetic gas*.[93–95] This syngas at high temperature and pressure[95] can be utilized for electricity and power generation, heat production and as a chemical feedstock for a variety of products such as ethylene, acetic acid, polyolefin, methanol, petrol and synthetic natural gas.[95-97] Syngas contains H_2, CH_4, CO and CO_2, and typically has a calorific value of 3.6–5.0 $MJ\,m^{-3}$.[98] Specific site conditions and the type of injected gas affect the calorific value of syngas. For air-injected syngas the calorific value ranges from 4.0 to 5.5 $MJ\,m^{-3}$, doubling with the injection of oxygen instead of air.[99]

UCG provides several economic and environmental benefits,[100] *e.g.* lower capital cost, as there is no need for a surface gasifier. Other benefits include the absence of coal mining infrastructure and of coal transportation and ash management costs, thus reducing the overall operating cost.[95] The need for underground coal mining may greatly reduce, thus enhancing coal utilization without increasing demand for labour and expensive equipment required for coal handling and transportation. UCG can enhance workable coal reserves as it can be applied to deposits which are not economically mineable by conventional methods, *e.g.* coals having low calorific values, thin seams, steeply inclined and thick deep seams.[101]

2.3.2 Methane Drainage and Utilization. Methane drainage is not itself a coal mining method; rather, it is a technique that helps in mining coal by reducing the amount of methane in-flow into mine workings. In this technique, boreholes, drainage pipes and/or openings are made in the coal seam before, during or after mining.[102,103]

Methane drainage, when carried out before starting coal mining activity, is used for decontamination of coal seams and aims to release as much methane as possible ahead of mining the coal seam. This methane from the coal seam is drained using a network of vertical, cross-measure and horizontal boreholes. Cross measure boreholes are drilled from an existing opening, such as a roadway, into the coal seam either in upward or downward direction in order to drain methane. The collected methane can be used as an energy resource, depending upon its amount and quality. In this way, the methane does not escape into the mine workings and is not released into the atmosphere, thus decreasing the environmental footprint of coal mining.[104]

During mining operations, if the surrounding strata contain a large amount of methane this can be drained by drilling boreholes from the sides, back, openings and mine workings. The surrounding strata are decontaminated and the methane thus drained is made available for use through the gas distribution system.

Methane can also be drained from the mined-out sections of a mine when it is released by strata relaxation and stress redistribution.[105,106]

2.3.3 Environmental Impacts of Novel Methods. UCG and methane drainage can help in the reduction of the environmental footprint of underground coal mining. In UCG, coal is not brought to the surface and the gaseous emissions and solid contaminants related to coal mining, haulage, transportation, storage and usage remain underground. Thus

UCG is environmentally superior to other coal exploitation methods. During the process, ash and other heavy metals stay underground, thus eliminating the cost of ash management facilities.[94] Other benefits include reductions in dust, noise, surface water pollution, water utilization and surface disturbance. No recovery requirement for mine water; minimal surface utilization; and reduction in surface hazard liabilities after abandonment give an edge to this technology. Since UCG does not require mining, the environmental hazards associated with dirt handling and disposal; coal washing and fines disposal; coal stocking and transport; and methane emission are greatly reduced or eliminated, resulting in a smaller environmental footprint.[101]

UCG also creates some environmental problems, *e.g.* groundwater pollution and subsidence; atmospheric emissions; human impact (noise, increased traffic, dust); sulfur and NO_x releases to surface, along with product gas; and the need for a proper CO_2 management system.[94,95,101]

Methane drainage provides an excellent alternative to otherwise wasted resource. The methane that is drained and used before, during, and after mining reduces the amount that is otherwise emitted to the surface. Methane is 25 to 86 times more potent than CO_2 as a greenhouse gas,[107] and its usage, rather than flaring in the atmosphere, provides an energy source as well as a reduction in overall GHG emissions from the mine.

3   Coal Transportation and Utilization

After coal is mined and brought to the surface, it is transported either to the coal washing, cleaning or processing facility, or to industry for consumption. This transportation occurs by conveyors, trucks and coal carts when the transportation distance is short. For long distance transportation specially designed trains, known as coal cars, and barges are used.[108] In some instances, coal can be converted into a slurry by mixing with water and transported through designated pipelines.[109]

3.1   Coal Transportation Methods

Trains are the most common domestic transportation system used for coal. Rail transportation accounts for approximately 58% of overall domestic coal transport and nearly 70% of coal for electric power generation in the USA.[108] Coal accounted for 36.9% of rail tonnage and 17.2% of rail revenue in the USA in 2015.[110] Most coal in the USA is consumed at power plants. Historically, coal has dominated US electricity generation because it is a cost-effective fuel choice, and freight railroads are a big reason for that. Approximately 70% of the coal delivered to coal-fuelled power plants is delivered by rail.[111] In 2014 the average cost to transport coal by train in the USA was $21.24 *per* ton, while the average total cost of coal delivered by train was $45.89 *per* ton; making trains the cheapest of all coal transportation modes.[112]

Waterborne barges account for 14% of coal transportation.[113] Barges are large vessels that carry coal through rivers and waterways and are an excellent means of coal transportation. Usually coal is transported un-covered in open-top barges. Due to the decline in coal production in the USA, these barges also are used for transporting other materials, including grains and farm products.[114] However, to transport grains the barges are often covered. In 2013, around three-fourths of the 32.7 million tons shipped by barge through the Port of Pittsburgh consisted of coal, lignite and coke.[115] Barge transportation costs around $5.96 *per* ton of coal in 2014 and $57.57 was their average delivered cost.[112]

Trucks are another important mode of coal transportation. Trucks trans-port 10% of the US coal used for electricity generation. In 2004, more than 85% of coal shipments were delivered to consumers by rail (684 million tons), road (129 million tons) or water (98 million tons).[116]

Conveyor belts are another mode of coal transportation but their use is limited to transportation of coal to the in-pit crushers or coal washing/pro-cessing facility located within the mine site.

3.2 Coal Utilization

The largest consumer of coal is the electricity-generation sector. In 2012, electricity generation accounted for 59% of world coal consumption and this trend continued at the same level until 2040.[117] The industrial sector utilized about 36% of world coal and the residential and commercial sector 4%.[117] In the USA, 33.2% of electricity was generated using coal as a fuel in 2015, with the projections for electricity generation being 33.4% in 2016 and also in 2017.[118] The electric power sector consumed about 92.5% of the total US coal consumption in 2014.[16]

3.2.1 Economic Impacts of Coal Transportation and Utilization. The direct impacts of coal transportation and utilization include the creation of jobs and an increase in tax revenue. With increased employment opportunities, the unemployment rate decreases and the overall economy improves with increased purchasing power.[119] The revenues of States increase with in-creased tax and royalty collections. In addition to that, coal provides the cheapest source of energy for the industrial and electricity-generation sec-tors. This, in turn, reduces prices of energy, consumable goods and com-mon household items. With the increased revenue, increased investment is made in infrastructure that provides an indirect boost to the economy. Indirect economic benefits of coal consumption and transportation are the development of associated industry, *e.g.* transportation hubs, loading docks and loading stations, *etc.*

Negative economic impacts of coal transportation include: traffic inter-ruption and delays; deterioration of roads; traffic congestion; decrease in property values; and impacts of noise on businesses and residential areas.[120]

3.2.2 Environmental Impacts of Coal Transportation and Utilization. En-vironmental impacts of coal transportation and utilization include: increased GHG emissions; coal dust and diesel exhaust; increased traffic and consumption of liquid fuel; air pollution; noise pollution; and increased accidents and fatalities.[121] Electricity generation using coal is regarded as one of the largest contributors of GHG emissions, especially carbon dioxide emissions. In 2015, CO_2 emissions from the electric power sector accounted for 37% of total US CO_2 emissions.[122]

4 Current Status of Coal Mining

Currently coal demand is in decline and the coal-mining sector is under tremendous pressure because of stricter regulations for coal-fired power generation. In 2015, coal demand dropped both in China and the USA for the first time in this century. Worldwide, coal-fired power plants have declined significantly, but China is building new coal-fired power plants in many countries.[123] More and more coal production is shifting from North America and Europe to Asia, where China and India are the dominant coal consumers and producers.[123] Coal is experiencing increasingly tough competition from natural gas (including shale gas) and renewables (wind and solar) but still is the largest fuel source for energy production.[57,117]

5 Future Trends

Although there is a decline in future coal demands, the world is currently burning more coal than was ever burnt in the past. Coal will continue to be a preferred source for power generation, but its share will decline to 36% by 2021.[124] From 2030 through 2040, coal will become the third-largest energy source, lagging behind liquid fuels and natural gas.[117]

The current US administration is pro-coal and is enacting policies that can help the revival of the coal industry, including relaxation of stream-dumping rules; rolling back the Obama administration's restrictions on carbon emissions from existing power plants; and blocking changes on coal-mining royalties.[125–127]

The push for clean coal technologies; the development of new technologies for using coal with a reduced environmental footprint; carbon capture and sequestration; installation of advanced air pollution controls; increasing consumption; production and prices of coal; shifting of coal markets; and, most importantly, the awareness and resolve of the coal mining industry to be environmentally responsible, is likely to continue a trend of increased coal use and increased coal-mining activities worldwide until 2040.[117,118,124]

References

1. C. E. Gregory, *A Concise History of Mining*, Revised edn, CRC Press, 2001.
2. U.S. Energy Information Administration (EIA), *International Energy Outlook 2016*, May 2016.
3. H. Xiang, Y. Kuang and C. Li, Impact of the China–Australia FTA on Global Coal Production and Trade, *J. Policy Model.*, 2017, **39**, 65–78.

4. IEA, *Key Coal Trends, Excerpts from Coal Information*, International Energy Agency, 2016.
5. World Coal Association, *Basic Coal Facts*, 2017. Retrieved January 07 2017, from https://www.worldcoal.org/basic-coal-facts.
6. D. F. Crickmer and D. A. Zegeer, *Elements of Practical Coal Mining*, Society of Mining Engineers of the American Institute of Mining, Metallurgical, and Petroleum Engineers, New York, 1981.
7. C. E. Gregory, *A Concise History of Mining*, A.A. Balkema, Lisse, The Netherlands, Exton, PA, 2001.
8. M. Lynch, *Mining in World History*, Reaktion Books, London, 2002.
9. H. J. Falcon-Lang, Earliest History of Coal Mining and Grindstone Quarrying at Joggins, Nova Scotia, and Its Implications for the Meaning of the Place Name "Joggins", *Atlantic Geol.*, 2009, **45**, 1–20.
10. American Coal Foundation, *Timeline of Coal in the United States*, 2005.
11. P. Darling, *SME Mining Engineering Handbook (3rd Edition)*, Society for Mining, Metallurgy, and Exploration (SME), 2011.
12. P. G. Marinos, *Engineering Geology and the Environment*, CRC Press, 1997.
13. H. L. Hartman and J. M. Mutmansky, *Introductory Mining Engineering*, John Wiley & Sons, 2002.
14. A. Rustan, C. Cunningham, W. Fourney, K. Simha and A. T. Spathis, *Mining and Rock Construction Technology Desk Reference: Rock Mechanics, Drilling & Blasting*, CRC Press, 2010.
15. World Coal Association, *Coal Mining*, 2017. Retrieved January 08 2017, from https://www.worldcoal.org/coal/coal-mining.
16. U.S. Energy Information Administration (EIA), *Annual Coal Report 2015*, 2016.
17. R. F. Munn, *Strip Mining: An Annotated Bibliography*, West Virginia University Library, Morgantown, 1973.
18. C. G. Treworgy, C. P. Korose, C. A. Chenoweth and D. L. North, *Availability of the Springfield Coal for Mining in Illinois, Department of Natural Resources, Illinois State Geological Survey*, 1999.
19. W. A. Hustrulid, R. K. Martin and M. Kuchta, *Open Pit Mine Planning & Design*, CRC Press, Boca Raton, London, 2013.
20. A. Akbari, M. Osanloo and M. Shirazi, Determination of Ultimate Pit Limits in Open Mines Using Real Option Approach, *IUST Int. J. Eng. Sci.*, 2008, **19**, 23–38.
21. M. Deutsch, E. Gonzalez and M. Williams, Using Simulation to Quantify Uncertainty in Ultimate-Pit Limits and Inform Infrastructure Placement, *Min Eng.*, 2015, **67**, 49–55.
22. R. A. Turpin, *A Technique for Assessing Ultimate Open-Pit Mine Limits in a Changing Economic Environment, Theses and Dissertations*, Lehigh University, 1980.
23. R. Tatiya, *Surface and Underground Excavations: Methods, Techniques and Equipment*, CRC Press/Balkema, Boca Raton, 2013.
24. T. Kose, Economic Evaluation of Optimum Bench Height In Quarries, *J. South. Afr Inst. Min. Metall.*, 2005, **105**, 127–135.

25. H. Soltanmohammadi, M. Osanloo, A. Sami and S. B. Malekzadeh, Selection of Practical Bench Height in Open Pit Mining Using a Multi-Criteria Decision Making Solution, *J. Geol. Min. Res.*, 2010, **2**, 48–59.

26. W. Gibson, I. Bruyn and D. Walter, *Considerations in the Optimisation of Bench Face Angle and Berm Width Geometries for Open Pit Mines*, International Symposium on Stability of Rock Slopes in Open Pit Mining and Civil Engineering, The South African Institute of Mining and Metallurgy, Johannesburg, 2006.

27. R. Elam, E. Teaster and M. Lawless, *Haul Road Inspection Handbook, MSHA Handbook Series, Handbook Number PH99-I-4, US Department of Labor*, 1999.

28. W. W. Kaufman and J. C. Ault, *Design of Surface Mine Haulage Roads: A Manual*, United States. Department of the Interior, Bureau of Mines., 2001.

29. P. Strohmayer, *Soil Stockpiling for Reclamation and Restoration activities after Mining and Construction, Department of Horticultural Science, University of Minnesota, Student Online Journal*, 1999, **4**, 1–6.

30. The Sustainable Business Review, *Top 5 Coal Mines in the World, 20-22 Wenlock Road, London. N1 7GU UK* 2014.

31. H. Krutka and L. Jingfeng, *Case Studies of Successfully Reclaimed Mining Sites, Cornerstone: The Official Journal of the World Coal Industry*, 2013.

32. R. Nel and M. S. Kizil, *The Economics of Extended Pre-Strip Stripping, 13th Coal Operators' Conference, University of Wollongong*, 2013.

33. J. Yamatomi and S. Okubo, *Surface Mining. Methods and Equipment*, Encyclopedia of Life Support Systems (EOLSS), 2011.

34. C. Latimer, *The Largest Machines in Mining*, Australian Mining, 2014. Retrieved February 02 2017, from https://www.australianmining.com.au/features/the-largest-machines-in-mining/.

35. E. A. Elevatorski, *Strip-Mineable Coals Guidebook*, Minobras, [Dana Point, Calif.] (P.O. Box 262, Dana Point 92629), 1980.

36. R. Mitra and S. Saydam, *Surface Coal Mining Methods in Australia*, INTECH Open Access Publishers, 2012.

37. A. Salama, J. Greberg and A. Gustafson, Luleå Tekniska Universitet, 2014.

38. M. M. Hansen, L. Kleinman and L. Vicklund, *Handbook of western reclamation techniques, Section 4 Top Soil, Office of Technology Transfer, Western Regional Coordinating Center, Office of Surface Mining Reclamation and Enforcement, Denver, CO*, 1996.

39. P. Hem, G. Fenrick and J. Caldwell, *Shovels*, InfoMine, 2012. Retrieved February 12 2017, from http://technology.infomine.com/reviews/Shovels/welcome.asp?view=full.

40. C. J. Bise, *Modern American Coal Mining: Methods and Applications*, Society for Mining, Metallurgy, and Exploration, Inc, Englewood, Colorado, 2013.

41. L. B. Paterson, Draglines - The Same "Old" Stuff?, International Mining Congress and Exhibition of Turkey-IMCET 2003, 2003.

42. B. A. Kennedy, M. Society for Mining and Exploration, *Surface Mining*, Society for Mining, Metallurgy, and Exploration, Littleton, Colo, 1990.
43. J. W. Martin and C. Martin, *Surface Mining Equipment*, Martin Consultants, Golden, Colo, 1982.
44. R. S. McGonigal, *Grades and CURVES, Railroading's Weapons in the Battle Against Gravity and GEOGRAPHY, Trains*, 2006.
45. R. Dawson, P. Winters and J. Matson, *Mountain Top Removal, Uhuru*, 2015, **10**, 12.
46. W. T. Plass, *History of Surface Mining Reclamation and Associated Legislation, Reclamation of Drastically Disturbed Lands*, 2000, 1–20.
47. F. Kulczak, *Illustrated Surface Mining Methods*, Skelly and Loy, New York, N.Y; Harrisburg, Pa, 1979.
48. Bureau of Mining Programs, *Auger Mining*, Document Number: 563-2112-604, Department of Environmental Protection, 1998.
49. M. Lukhele, Surface Auger Mining at Rietspruit Mine Services (Pty) Ltd, *J. South. Afr. Inst. Min. Metall.*, 2002, **102**, 115–119.
50. R. McKinney, *MSHA Procedures for Assigning Legal Identity Numbers and Relocation Notices at Auger and Highwall Mining Operations*, Coal Mine Safety and Health Administration, 2004.
51. V. Bogdetsky, K. Ibraev and J. Abdyrakhmanova, *Mining Industry as a Source of Economic Growth in Kyrgyzstan, World Bank, Bishkek.* © *World Bank.* https://openknowledge.worldbank.org/handle/10986/16959, *License: CC BY 3.0 IGO*, 2005.
52. I. Dorin, C. Diaconescu and D. I. Topor, The Role of Mining in National Economies, *Int. J. Acad. Res. Acc., Finance Manage. Sci.*, 2014, **4**, 155–160.
53. G. Hilson, *Small Scale Mining and its Socio Economic Impact in Developing Countries*, Natural Resources Forum, 2002.
54. U.S. Energy Information Administration (EIA), *Average Number of Employees at Underground and Surface Mines by State and Mine Production Range, 2015, Annual Coal Report 2016*, 2016.
55. J. Rolfe, B. Miles, S. Lockie and G. Ivanova, Lessons from the Social and Economic Impacts of the Mining Boom in the Bowen Basin 2004-2006, *Aust. J. Reg. Stud.*, 2007, **13**, 134.
56. M. Matheis, Local Economic Impacts of Coal Mining in the United States 1870 to 1970, *J. Econ. Hist.*, 2016, **76**, 1152–1181.
57. IEA, *Renewable Electricity Generation Climbs to Second Place after Coal, International Energy Agency*, 2015.
58. K. Spitz and J. Trudinger, *Mining and the Environment: From Ore to Metal*, CRC Press, Boca Raton, 2009.
59. A. G. N. Kitula, The Environmental and Socio-Economic Impacts of Mining on Local Livelihoods in Tanzania: A Case Study of Geita District, *J. Cleaner Prod.*, 2006, **14**, 405–414.
60. C. L. Carlson and D. C. Adriano, Environmental Impacts of Coal Combustion Residues, *J. Environ. Qual.*, 1993, **22**, 227–247.
61. R. Tiwary, Environmental Impact of Coal Mining on Water Regime and Its Management, *Water, Air, Soil Pollut.*, 2001, **132**, 185–199.

62. W. J. Rankin, *Minerals, Metals and Sustainability: Meeting Future Material Needs*, CSIRO Publishing, Collingwood, Vic, 2011.
63. D. B. Johnson and K. B. Hallberg, Acid Mine Drainage Remediation Options: A Review, *Sci. Total Environ.*, 2005, **338**, 3–14.
64. C. A. C. Iii and M. K. Trahan, Limestone Drains to Increase PH and Remove Dissolved Metals from Acidic Mine Drainage, *Appl. Geochem.*, 1999, **14**, 581–606.
65. D. A. Kepler and E. C. McCleary, *Successive Alkalinity-Producing Systems (SAPS) for the Treatment of Acidic Mine Drainage, Bureau of Mines Special Publication SP 06B-94*, 1994, **1**, 185–194.
66. K. B. Hallberg and D. B. Johnson, Novel Acidophiles Isolated from Moderately Acidic Mine Drainage Waters, *Hydrometallurgy*, 2003, **71**, 139–148.
67. A. Akcil and S. Koldas, Acid Mine Drainage (AMD): Causes, Treatment and Case Studies, *J. Cleaner Prod.*, 2006, **14**, 1139–1145.
68. C. A. Cravotta, Size and Performance of Anoxic Limestone Drains to Neutralize Acidic Mine Drainage, *J. Environ. Qual.*, 2003, **32**, 1277–1289.
69. D. Blowes, C. Ptacek, J. Jambor and C. Weisener, The Geochemistry of Acid Mine Drainage, *Treat. Geochem.*, 2003, **9**, 612.
70. United States Congress. Office of Technology Assessment, *Western Surface Mine Permitting and Reclamation*, Congress of the U.S., Office of Technology Assessment, Washington, D.C, 1986.
71. K. M. Wantzen and J. H. Mol, Soil Erosion from Agriculture and Mining: A Threat to Tropical Stream Ecosystems, *Agriculture*, 2013, **3**, 660–683.
72. United States. Department of the Interior, *Surface Mining and our Environment: A Special Report to the Nation*, Washington, 1967.
73. S. M. Rupprecht, *Mine Development–Access To Deposit, University of Johannesburg*, 2012, 101–121.
74. H. L. Hartman and J. M. Mutmansky, *Introductory Mining Engineering*, J. Wiley, Hoboken, N.J, 2002.
75. EMFI, *Coal Mining Methods, Energy and Mineral Field Institute, Colorado School of Mines*, 1994.
76. S. S. Peng, *Coal Mine Ground Control*, Wiley, New York, 1978.
77. M. M. Murphy, Shale Failure Mechanics and Intervention Measures in Underground Coal Mines: Results From 50 Years of Ground Control Safety Research, *Rock Mech. Rock Eng.*, 2016, **49**, 661–671.
78. A. T. Iannacchione and S. C. Tadolini, *Coal Mine Burst Prevention Controls*, 27th International Conference on Ground Control in Mining, 2008.
79. K. Noack, Control of Gas Emissions in Underground Coal Mines, *Int. J. Coal Geol.*, 1998, **35**, 57–82.
80. S. S. Peng, *Longwall Mining*, Department of Mining Engineering, West Virginia University, Morgantown, WV, 2006.
81. K. S. Stout, *Mining Methods & Equipment, Mining Informational Services*, McGraw-Hill, New York, N.Y, 1980.

82. J. Clark, J. H. Caldon and E. A. Curth, *Thin Seam Coal Mining Technology*, Noyes Data Corp, Park Ridge, N.J, 1982.

83. Kentucky Geological Survey, *Methods of Mining, University of Kentucky*, 2012.

84. C. Mark, *Deep Cover Pillar Recovery in the US*, Proceedings of the 28th International Conference on Ground Control in Mining, 2009.

85. F. E. Chase, C. Mark and K. A. Heasley, *Deep Cover Pillar Extraction in the US Coalfields*, 21st International Conference on Ground Control in Mining, Morgantown, 2002.

86. K. Heasley, *The Forgotten Denominator, Pillar Loading*, 4th North American Rock Mechanics Symposium, 2000.

87. C. Mark, F. E. Chase and A. A. Campoli, *Analysis of Retreat Mining Pillar Stability*, West Virginia Univ., Morgantown, WV (United States), 1995.

88. C. Mark, F. Chase and D. Pappas, Reducing the Risk of Ground Falls during Pillar Recovery, *Trans. Soc. Min. Metall. Explor. Inc.*, 2003, **314**, 153–160.

89. E. Ghasemi, M. Ataei, K. Shahriar, F. Sereshki, S. E. Jalali and A. Ramazanzadeh, Assessment of Roof Fall Risk During Retreat Mining in Room and Pillar Coal Mines, *Int. J. Rock Mech. Min. Sci.*, 2012, **54**, 80–89.

90. R. K. Zipf, *Toward Pillar Design to Prevent Collapse of Room and Pillar Mines*, 108th Annual Exhibit and Meeting, Society for Mining, Metallurgy and Exploration, 2001.

91. Ö. Uysal and A. Demirci, Shortwall Stoping Versus Sub-Level Longwall Caving-Retreat in Eli Coal Fieds, *J. South Afr. Ins. Min. Metall.*, 2006, **106**, 425.

92. H. Beynon, A. W. Cox and R. Hudson, *Digging Up Trouble: The Environment, Protest and Opencast Coal Mining*, Rivers Oram, New York; London, 2000.

93. P. N. Thompson, J. R. Mann and F. Williams, *Underground Coal Gasification*, National Coal Board, London, 1976.

94. K. J. Fergusson, A Cleaner, Cheaper, Indigenous Fuel for Combined Cycle Plants, *Modern Power Syst.*, 2009, **29**, 24–26.

95. E. Burton, J. Friedmann and R. Upadhye, in *Draft Technical Report (Lawrence Livermore National Laboratory, U.S. Department of Energy contract No. W-7405-Eng-48.)*, 2006, p. 119.

96. O. B. E. Rohan Courtney, *UCG Workshop*, Pittsburgh Coal Conference, Pittsburgh, PA, 2009.

97. J. Liu, C. Mallet, A. Beath, D. Elsworth and B. Brady, in *GeoProc*, Sweden, 2003.

98. M. K. Ghose and B. Paul, Underground Coal Gasification: A Neglected Option, *Int. J. Environ. Stud.*, 2007, **64**, 777–783.

99. L. Walker, Underground Coal Gasification: A Clean Coal Technology Ready for Development, *Aust. Coal Rev.*, 1999, 19–21.

100. R. A. Meany and A. Maynard, *A Review of the Potential for Underground Coal Gasification and Gas to Liquids Applications in Pedirka Basin*,

Onshore Northern Territory and Pela 77 Pedirka basin, Onshore South Australia Mulready Consulting Services Pty Ltd, 2009.

101. D. P. Creedy, K. Garner, S. Holloway, N. Jones and T. X. Ren, *Review of Underground Coal Gasification Technological Advancements, Report No. COAL R211, DTI/Pub URN 01/1041 Under Department of Trade & Industry, UK*, August 2001.

102. N. Szlazak, D. Obracaj and J. Swolkien, Methane Drainage from Roof Strata Using an Overlying Drainage Gallery, *Int. J. Coal Geol.*, 2014, **136**, 99–115.

103. Y. Tang, Methane Drainage Optimization by Roof-Borehole Based on Physical Simulation, *Arab. J. Geosci*, 2015, **8**, 7879–7886.

104. H. Zhou, H. Dai and C. Ge, Quality and Quantity of Pre-Drainage Methane and Responding Strategies in Chinese Outburst Coal Mines, *Arab. J. Geosci.*, 2016, **9**, 1–14.

105. W. Qin, J. L. Xu and G. Z. Hu, Numerical Simulation of Abandoned Gob Methane Drainage through Surface Vertical Wells, *PLos One*, 2015, **10**, 18.

106. W. Qin, J. Xu and G. Hu, Optimization of Abandoned Gob Methane Drainage Through Well Placement Selection, *J. Nat. Gas Sci. Eng.*, 2015, **25**, 148–158.

107. Z. Hyder, N. S. Ripepi and M. E. Karmis, A Life Cycle Comparison of Greenhouse Emissions for Power Generation from Coal Mining and Underground Coal Gasification, *Mitigation Adapt. Strat. Global Change*, 2016, **21**, 515–546.

108. EIA, *Coal Mining and Transportation, Coal Explained*, 2016.

109. WCA, *Coal Market & Pricing*, World Coal Association, 2017. Retrieved January 21 2017, from https://www.worldcoal.org/coal/coal-market-pricing.

110. Association of American Rail Road, *Class I Railroad Statistics*, 2016.

111. Association of American Rail Roads, *Railroads and Coal*, 2016.

112. EIA, *Real Average Transportation and Delivered Costs of Coal*, U.S. Energy Information Administration (EIA), 2016.

113. A. Caplan, *Waterborne Logistics: The Shift in Coal Barge Demand, Argus*, 2016.

114. A. Moore, *Us Coal Barge Prices Down Substantially Due to Lack of Coal Traffic, S&P Global Platts*, 2014.

115. A. Aupperlee, *Content of Barges Diversified in Western PA. as Coal Taking up Less Space, TRIB Live*, 2015.

116. N. R. Council, *Coal: Research and Development to Support National Energy Policy*, National Academies Press, 2007.

117. U.S. Energy Information Administration (EIA), *International Energy Outlook*, 2016.

118. American Coalition for Cleaner Electricity (ACCCE), *Coal Facts*, 2016.

119. E. Thompson, *The Economic and Tax Revenue Impact of Coal Industry Activity in Nebraska: Final Report, Bureau of Business Research, Department of Economics*, 2014.

120. City of Seattle, *Economic Analysis of Proposed Coal Train Operations*, 2013.

121. M. J. Chadwick, N. Highton and N. Lindman, *Environmental Impacts of Coal Mining & Utilization: A Complete Revision of Environmental Implications of Expanded Coal Utilization*, Elsevier, 2013.

122. U.S. Energy Information Administration (EIA), *How much of U.S. Carbon Dioxide Emissions are Associated With Electricity Generation?*, 2016.

123. C. A. Fernandez, *IEA's Medium Term Coal Market Report 2016*, International Energy Agency, 2017.

124. IEA, *Coal, Medium-Term Market Report, Market Analysis and Forecasts to 2021*, International Energy Agency, 2016.

125. M. Brown, *Trump Administration Blocks Changes on Coal Mining Royalties, The Washington Times*, 2017.

126. D. Boyce, *Trump To Begin Rollback of Coal Regulations, Inside Energy*, 2017.

127. D. Walsh, *GOP House Votes to Reject Obama Administration Stream Protection Rule, CNN*, 2017.

Coal-fired Power Stations

LUCAS KRUITWAGEN,* SETH COLLINS AND BEN CALDECOTT

ABSTRACT

Thermal coal-fired power stations currently provide approximately 40% of the world's electricity and 30% of the world's generating capacity. Approximately 83% of all coal demand is thermal coal, and 61% of primary coal energy is consumed in power stations. Notwithstanding alternatives in coal gasification, coal-to-liquids, and chemical looping technologies, the future of coal in the 21st century depends largely on the future of coal combustion for power generation. This chapter provides a technical overview of coal-fired power stations and their exposure to a wide array of environment-related risks, including greenhouse gas emissions and stranded assets; water consumption and competition with agriculture, industry, and domestic uses; climate stresses induced by anthropogenic climate change (of which they are the primary cause); competition with renewables and generating flexibility; costs and trade-offs of mitigation options; retrofitability with carbon capture and storage; and the availability of finance. The future of coal in the 21st century depends largely on the response of policy makers, industry and the concerned public to these risks.

1 Introduction

The future of coal in the 21st century depends largely on the future of coal-fired power. Thermal coal-fired power stations currently provide

---

*Corresponding author.

Issues in Environmental Science and Technology No. 45
Coal in the 21st Century: Energy Needs, Chemicals and Environmental Controls
Edited by R.E. Hester and R.M. Harrison
© The Royal Society of Chemistry 2018
Published by the Royal Society of Chemistry, www.rsc.org

approximately 40% of the world's electricity and 30% of the world's generating capacity.[1] Approximately 83% of all coal demand is thermal coal,[2] and 61% of primary coal energy is consumed in power stations.[1]

Coal-fired power stations produce approximately 31% of global CO_2 emissions.[1] The policies and technologies required for the transition to a zero-carbon energy system is a fundamental threat to the companies owning and operating coal-fired power stations. As large fixed assets with long economic lifespans, coal-fired power stations are particularly at risk of becoming stranded assets: assets which have suffered unanticipated or premature write-downs, devaluations, or conversion to liabilities. The tension between the transition risks of the decarbonisation of the energy system and the risks associated with catastrophic climate change has never been higher.

A more inclusive and sustainable global economy will necessitate a steep drop in coal-fired power generation. The International Energy Agency's (IEA) 450 Scenario (a scenario calibrated to have a 50% chance of constraining warming to 2 °C)[3] projects global coal-fired power dropping to 7% of generation and 10% of generating capacity by 2040, relative to 28% of generation and 22% of generating capacity in the central New Policies Scenario and 40% of generation and 30% of generating capacity today.[1] In the 450 Scenario, 70% of coal-fired electricity would be generated from Carbon Capture and Storage (CCS)-equipped power stations, requiring a global coal-fired CCS fleet of about 260 GW (up from less than 200 MW today).[4] While the 450 Scenario projects substantial investment in new fossil-fuel power stations (particularly high efficiency, low-emissions technologies, which are discussed in this chapter), these new installations come at the expense of the existing generating stock, 60% of which do not reach the end of their technical lifespan, failing to fulfil their expected investment returns. A further 8% of this subset fail to recover even their investment costs.[5] These losses may now already be unavoidable; the 'locked-in' capital stock of coal-fired generating stations may exceed the limit compatible with a 2 °C warming future as early as 2017.[6]

Findings from the International Panel on Climate Change's Fifth Assessment Report Working Group III (IPCC's AR5, WGIII)[7] provide a similar outlook. By 2050, the electricity system must be already emissions netnegative under cost-efficient projections constrained to 450 ppm CO_2 equivalent (CO_2eq). Projections restricted to not use CCS feature even more rapid electricity system decarbonisation and extremely net-negative Agriculture, Forestry and Other Land Uses (AFOLU) change.[7] CCS-limited projections also impose higher mitigation costs than unconstrained options.[7] IPCC's AR5, WGIII are not explicit about the role of coal-fired power in the generating mix nor the extent to which existing power stations might be stranded; however, they do indicate that investment levels in fossil-fuel power generation until 2030 must reduce by approximately $66bn.[7]

Few comprehensive projections are available that correspond to a warming limit of 'likely' less than 1.5 °C. The metastudy of Rogelj *et al.*[8] collects evidence that the transition to a 1.5 °C-warming compatible pathway requires the immediate and substantial decarbonisation of electricity generation, beyond what is already required for 2 °C-compatibility, and faster than both transportation and industrial decarbonisation.

Across scenarios and outlooks, decarbonising electricity generation early and rapidly is the highest priority for constraining climate change to 2 °C or less while minimising mitigation costs.[1,7,8] Achieving a 2 °C constraint to warming will strand both existing and new-build power stations. Pathways are sensitive to the availability of CCS as a measure to reduce the emissions of new-build stations and avoid stranding existing power stations. Pathways without CCS available or that achieve 1.5 °C warming rely on even faster decarbonisation of the global generating mix. As CCS is considered extensively by Anthony in Chapter 7 of this volume,[9] our treatment of CCS later in this chapter will be argumentative rather than illustrative.

Even without large-scale policy action to limit greenhouse gas emissions, the future of coal-fired power is far from certain. The falling cost of renewables and availability of shale gas is eliminating coal's cost advantage in many regions around the world.[10] Concerns over air, land and water pollution, and competition over water resources with agriculture or domestic uses may create policy environments which are unfavourable for coal development.

The massive expansion of coal for power generation at the start of the 21st century has recently stalled, mainly driven by changing policies and economic conditions in China and India, shown by overall drops across the thermal power generation stock in pre-construction activity (48%), construction starts (62%) and ongoing construction (19%). Globally, more construction is now frozen than has entered the development pipeline.[11] Coal will either be the first victim of the transition to a sustainable world prosperously enjoyed by future generations or it will be the salient perpetrator of climate change-induced sea-level rise; property destruction; extreme weather severity and frequency; drought and floods; crop failure; mass migration; biodiversity loss and heat stress; loss of life and economic productivity caused by air pollution; and conflict over scarce water resources.

Sections 2 and 3 describe technologies for thermal coal preparation and combustion, some of which have been the subject of recent attention due to their improved efficiency, reduced emissions, or compatibility with CCS. Section 4 presents an analysis of the environment-related risks which will dominate the fortunes of coal-fired power stations in the 21st century. Section 5 concludes.

2   Pre-treatment for Power Generation

Pre-treatment of coal is performed on 'run of mine' (ROM) coal before use in power stations or sale on coal markets. Coal pre-treatment can be used to control greenhouse gas or conventional air pollution emissions or to provide compatibility with certain types of emerging combustion technologies. As coal markets, and energy markets more broadly, change the technical specifications of the products they demand, coal pre-treatment is required to adjust the petrographic properties of coal including its energy, sulfur, moisture, volatile organic, and ash content. These changes will have important implications for the companies and countries that supply the world's coal.

2.1 Conventional

2.1.1 Technology Description. Also called 'beneficiation' or 'washing', conventional coal preparation reduces coal particle size, removes impurities to reduce pollution, and improves the heating value of coal. A general description of coal preparation includes various crushing and screening stages, volumetric and gravitational separation methods, and dewatering prior to final preparation for transport. A brief overview is given here; see Leonard and Hardinge,[12] Nunes,[13] or Kilma *et al.*[14] for details on methods.

Marketed coal is technically specified by its energy density and its impurities, including sulfur, moisture, volatile organic compounds and ash content. Ash is a generalised term for the non-organic content in coal which cannot be combusted – the rock intermixed in run-of-mine (ROM) coal. Conventional coal pre-treatment seeks to remove these impurities using a variety of methods. A generalised flow diagram for coal pre-treatment is shown in Figure 1.

As an initial step, ROM coal is crushed and sized. Crushing reduces coal and contaminant particle sizes for subsequent processes and eventual combustion. Common technologies are roller crushers, rotary breakers and hammer mills, some of which are also designed to reject rock particles based on hardness. Crushed coal is then sized according to the maximum and minimum size of individual particles using screen and sieves. Coal is saturated into a slurry to 'deslime' the feedstock of fine clays and soluble

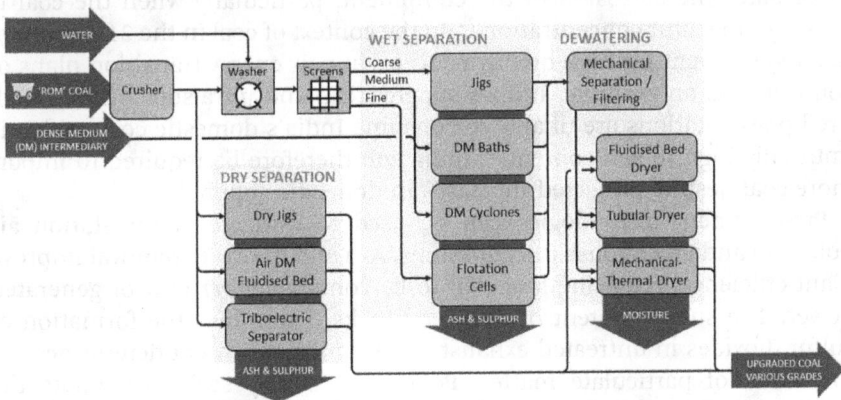

Figure 1 Generalised flow diagram for coal pre-treatment.

contaminants. Size-sorted coal is categorised broadly on a spectrum from 'coarse' to 'fine' as it moves on to further pre-treatment steps.

Sulfur and ash contaminants are denser than pure organic coal, so most separation processes use gravitational methods to remove more-dense particles. Coal and ash slurries are sometimes mixed with an intermediate density medium, which assists the separation of low-gravity coal and high-gravity ash. The coal slurry then proceeds to a gravitational separation technology. Technologies used to separate coarse coal slurries include jigs, trough, drum and cone-type vessels. Medium and fine coal slurries use additional cyclone technologies and fine coal slurries may be finally separated using advanced floatation techniques.

Cleaned coal slurries must finally be dewatered to prevent moisture inefficiencies in combustion, reduce shipping weights, and avoid undue water consumption. Mechanical sedimentation and filtration techniques are used to remove as much free water as possible. Final stage heating removes chemically bound water to the extent desired for the end product. Water and dense medium intermediary materials are recycled as well as possible, and removed contaminants are sent to landfill. There are some coal pretreatment techniques which do not require the formation of a coal slurry. Electrostatic, magnetic, and pneumatic dry-sorting techniques are being developed; however, they have not yet had significant uptake by industry.

2.1.2 Conventional Pre-treatment in the 21st Century. Even conventional coal preparation technologies have a critical role to play for coal in the 21st century. Climate and conventional pollution concerns, changing power markets, and even concerns over energy independence will change the types and prices of coal desired by markets and end-users.

Advanced combustion technologies like super- and ultra-supercritical steam generators require a higher quality coal than subcritical boilers. At the higher temperatures of supercritical steam generators, ash deposits on internal heat-transfer surfaces are in the liquid phase. The liquid phase ash exacerbates the corrosion of the equipment, particularly when the coal is also high in sulfur concentration.[15] In the context of coal in the 21st century, this requirement for high-quality coal can impair energy-transition plans of countries. As an example, India's supercritical and ultra-supercritical coal-fired power stations are unable to consume India's domestic coal, which is unusually high in ash content.[16] India will therefore be required to import more coal despite projected increases in domestic supply.[2]

Pre-treatment technologies can be used to mitigate power station air pollution and greenhouse gas emissions. Ash and moisture removal improve plant efficiency, requiring less coal to be combusted *per* unit of generated power. The sulfur content of coal feedstocks determines the formation of sulfur dioxides in untreated exhaust gases, and ash content determines the formation of particulate matter. Policymakers may seek to regulate the quality of coal as an indirect method of reducing power station conventional air-pollution emissions. As an example, China banned imports of high-sulfur, high-ash coal in 2015, citing air pollution concerns.[17] Coal exporters wishing to sell coal to consumers in China must now bear the cost of coal beneficiation. This regulation is likely to hurt Australian exporters the most, whose coal is high in ash content.

2.2 Gasification

2.2.1 Technology Description. Gasification for coal is the thermal decomposition of feedstock to produce carbon monoxide (CO) and hydrogen (H_2), as well as other gaseous compounds. This synthesis of gases is referred to as 'syngas'. The process is driven by the heat of the partial oxidisation of feedstock occurring in an oxygen-scarce environment. Syngas is a building block for a variety of applications, including oil refining, consumer products, natural gas substitutes, chemicals, fertilisers, transportation fuels, steam and power. See Higman and Fernando for general references on gasification and coal gasification, respectively.[18,19]

Broadly, the process of gasification is as follows: a feedstock—coal, biomass, petroleum coke or waste—is fed into a gasifier, which separates the feedstock into gaseous constituents and solids. The solid by-products are sold off, while the gaseous constituents are further processed through a gas stream clean-up and component separation targeted at eliminating particulates and sulfuric compounds.[20] The gas, now purely carbon monoxide and hydrogen, can then be used for a variety of applications, including fuels, chemicals and power, as in an Integrated Gasifier Combined Cycle power station described below. One such process is schematically shown in Figure 2.

Gasification for coal produces syngas by exposing pulverised coal to controlled amounts of air or pure oxygen under high temperatures and pressures. The syngas is then cooled down and purified. Steam may be added to the syngas to oxidise carbon monoxide and release hydrogen *via* the water gas shift reaction. This raw gas also contains many contaminants

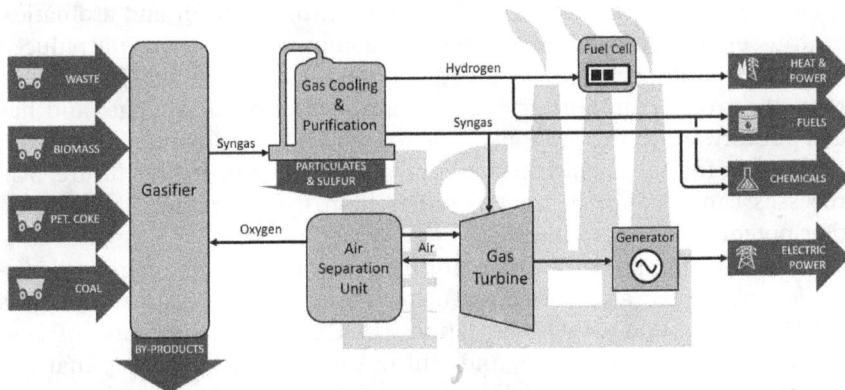

Figure 2   Gasification and energy conversion generalised flow diagram.

undesirable for further gas utilisation. These are removed through a process of cooling and then separation of dust contaminants, which condense out – for example, removing ammonia and hydrogen chloride through water wash. Sulfurous components must also be removed by conversion to sulfuric acid.

Coal feedstocks can be substituted by biomass. Biomass sources include switch grass, corn husks, lumbering and timbering wastes, yard wastes, construction and demolition wastes, and bio solids. Biomass often contains high concentrations of moisture, which reduce temperatures in the gasification chamber and thus lower efficiencies. Biomass plants are generally smaller than coal or petroleum coke gasification plants, and thus are less expensive to build. Biomass gasification of waste products has a strong potential to positively use traditional landfill waste productively, reducing costs and generating power or producing ethanol from non-food biomass sources. For example, in comparison to power generation from conventional mass-burn incineration, gasification of the same ton of municipal solid waste can be used to generate nearly twice the electricity. Gasification turns waste from a fuel to a feedstock, meaning that it is used to create higher value products beyond the capabilities of an incinerator, and allows better control of contaminants.

Coal gasification opens new options for coal to compete with liquid and gas energy carriers. While the interest of this chapter is to examine the future of coal in the generation of power, the synergy of gasification with coal-to-liquids (CTL) technologies is worth discussing. CTL enables coal to be converted to liquid fuels *via* two major methods – direct and indirect coal liquefaction (DCL and ICL respectively). The former, direct coal liquefaction, converts coal to liquid fuels under high heat, high pressure, and in the presence of a catalyst to initiate hydro-cracking. The latter, indirect coal liquefaction, converts coal to syngas and then syngas to liquid fuels *via* the Fischer–Tropsch process. CTL is covered at length by Snape in Chapter 6 of in this volume.[21]

DCL is commonly regarded as more efficient for production of liquid fuels, as it requires only partial breakdown of the coal. However, ICL fuels are cleaner, as they are essentially free from nitrogen, sulfur and aromatics, and thus emit fewer contaminants when combusted. As well as a reduced environmental impact, ICL has greater variability and flexibility in outcome products, stronger supporting infrastructure and past knowledge, and has been put forward as the more likely option for CTL development. Moreover, if hydrogen fuel cells gain importance and utilisation in the future, ICL processes can produce hydrogen, rather than hydrocarbons, creating another potential future application.

The gasification process, which normally occurs in an above-ground gasification plant, can also take place below ground in coal seams, a process called 'Underground Coal Gasification' (UGC). UCG is a process of converting coal that is unworked, and still in the ground, into a gas that can be utilised in power generation, industrial heating or manufacture of synthetic fuels.

The UCG process requires drilling two wells into the coal seam, which are then heated to a high temperature with oxidants injected through one well. Water is also needed and may be pumped from the surface or may come from the surrounding rock. The coalface is ignited and, at high temperatures (1500 K) and high pressures, this combustion generates carbon monoxide, carbon dioxide and hydrogen. Oxidants react with the coal to create syngas, which is then drawn out through the second well. Pressure is naturally occurring at seam depths, normally below 1200 ft, and the contaminants are left underground.

UCG projects have been active in Russia, the United States, Australia, China, India, and South Africa.[22,23] UCG has certain advantages over coal-to-liquid/coal-to-gas (CTL/CTG) in terms of lower plant costs, less surface emissions of sulfur and nitrous oxides, and potential synergies with CCS as CO_2 can be stored in coal cavities after gasification and extraction.[24] Key concerns include fugitive methane emissions, carbon intensity of products, water and land pollution near extraction sites, water intensity, and reputational risks.[22]

2.2.2 Gasification of Coal in the 21st Century. Recently, the worldwide gasification market has seen a surge of demand growth, driven by coal gasifiers. The economic arguments for gasification for power and products focus on its ability to increase the value of low-price or negative-value feedstocks through gasification.[25] Feedstock flexibility gives power stations in particular a real option to decarbonise later in their economic life by switching to low-carbon fuels like biomass. Major proponents focus on benefits such as reducing municipal solid waste, reducing dependence on imported natural gas, reduced unusable waste products (compared to incineration), the ability to produce multiple products of value, and the flexibility of feedstock inputs. However, both biomass and municipal waste are minimal feedstocks to the current operating fleet,[26] as shown in Figure 3. While the benefits of biomass and waste conversion are highly touted, their actual roles are negligible, mainly due to transportation costs, and future developments are almost primarily planned for coal (see Figure 4).[26]

As of 2014, 272 gasification plants were in operation, with 74 plants under construction. Regional concentration is dominated by Asia and Australia, followed by Africa, Europe, North America, and Central and South America, driven by growth in chemical, fertiliser and coal-to-liquids industries. The gasification market is dominated by chemicals, and developments in China have moved the production of gaseous fuels into the second-most used application, followed by liquid fuels and then, finally, power.

Gasifiers do not avoid entirely the environmental and economic challenges associated with coal energy conversion. With coal as a feedstock, many of the harmful pollutants and particles from conventional coal pretreatment and combustion are also a concern with coal gasification. Syngas can be directly treated for mercury and sulfur content prior to any further

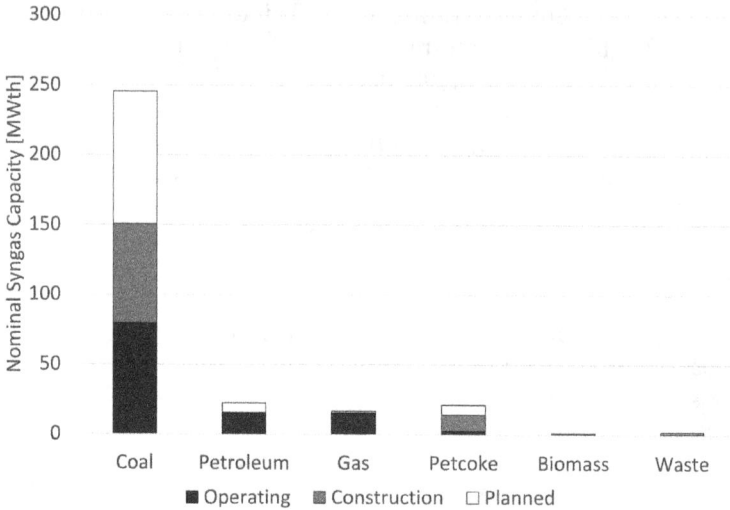

Figure 3   Global installed gasifier capacity by fuel.
(© Higman Consulting GmbH,[26] reproduced with permission.)

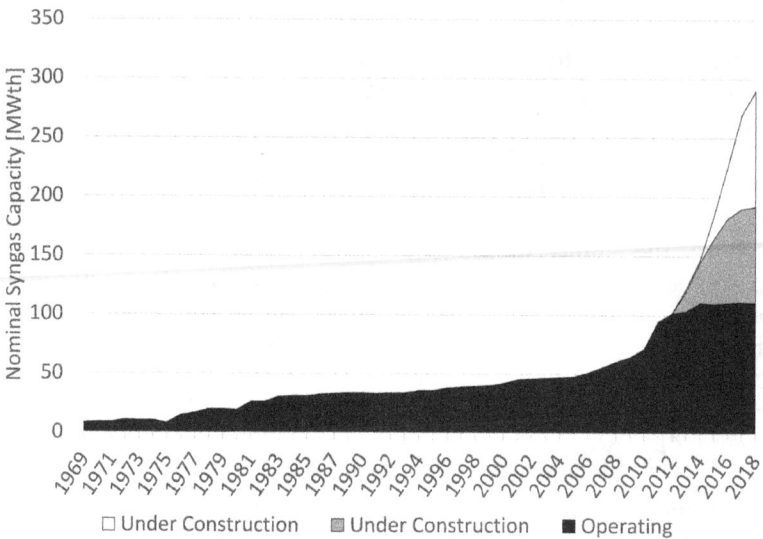

Figure 4   Global growth in gasifier capacity.
(© Higman Consulting GmbH,[26] reproduced with permission.)

processing or use. Gasification slag—made up of ash, unreacted carbon, and metals produced by the gasifier—is less expensive to remove than in combustion-based systems and is useable in making cement or asphalt.

For power generation, coal-derived synthetic natural gas (SNG) is rarely competitive with conventional natural gas. A comparative life-cycle cost analysis of conventional natural gas and synthetic gas shows that the environmental and societal costs of SNG production are much higher than that of natural gas.

Coal gasification as a feedstock for liquids production also faces considerable barriers. Very low thermal efficiencies are associated with hydrocarbon liquefaction – in the range of 45–55%. Given that substantial volumes of coal are required to generate fuels in any useful amount, these technologies are only viable in areas with abundant coal reserves, limiting large-scale production to six nations globally (Australia, China, India, South Korea, South Africa and the United States) and creating infrastructure issues.[27] Second, coal must be carefully quality-controlled for low sulfur to prevent denaturing of expensive catalysts. Third, CTL/CTG plants are very costly to build and construction takes four to five years. Moreover, since a long plant life is crucial to guaranteeing a return for investors, local coal reserves must be sufficient to ensure this is possible.

Coal gasification projects have made some progress in emerging markets, but in general have been hamstrung by delays. Table 1 shows headlines and content sampled from the quarterly newsletter *NETL Gasification and Transportation Fuels Quarterly News* since 2015. Many of the positive headlines concern waste to energy projects, misrepresenting how small a segment of the market they make up.[28]

Table 1   Summary of headlines in the NETL G&T Fuels Quarterly News. (Drawn from the National Energy Technology Laboratory.[28])

Headline	News type	
Duke Energy Will Foot $85M More in Coal Gasification Plant Costs under Indiana Settlement	Negative	Cost overrun
Southern Faces $234 Million Tax Bill Due to Kemper Delays	Negative	Cost overrun
Kemper Power Plant Overruns Climb another $25 Million	Negative	Cost overrun
Underground Coal Gasification Project put on Hold	Negative	Project hold
Economists Claim Coal Gasification 'Could Generate £13bn'	Positive	Economic value
Adani Group to Invest $3.75 Billion in Coal Gasification Project in Chhattisgarh	Positive	Economic value
Russia's Gazprom in Partnership Deal for South African Coal Gasification Project Diversified	Positive	Project plan
Coal Gasification Urea Plant to Come Up in Chhattisgarh, India	Positive	Project plan
€202M Contract for 15 MW Waste to Energy Gasification Plant in Northern Ireland	Positive	Waste project
£150m EPC Contract for 28 MW Waste to Energy Gasification Plant in Hull, UK	Positive	Waste project
HTDC Unveils New $6.8 Million Waste to Energy System	Positive	Waste project
Lockheed Martin Secures $43m EPC Contract for 5 MW Waste Gasification Plant in Germany	Positive	Waste project

Gasification as a pre-treatment technology of coal for feedstocks for liquids, fuels and other products face steep challenges with regard to costliness and efficiency. As a pre-treatment technology for power generation there is yet no dramatic uptake of the technology or pipeline of projects to suggest that coal-fuelled gasification will be a productive or sustainable source of power generation in the 21st century.

3   Combustion Technologies

3.1   Boilers and Steam Generators

Coal-fired boilers are the most common and conventional technology for recovering energy from thermal coal. Pulverised coal is injected into a chamber where it is combusted. The heat is used to generate steam, which drives a steam turbine, which in turn drives a generator. This section will consider the thermodynamic and equipment differences between conventional subcritical boilers and various classifications of supercritical steam generators. These advanced steam generators have higher thermal efficiencies and thus lower emissions, making them appealing options for coal combustion under constrained greenhouse gas and conventional air pollutant pathways. They are called steam generators because water does not undergo a conventional boiling process from liquid to gas, rather it exists as a critical fluid. The purpose of this section is to comment on the difference between conventional boiler technologies and advanced steam generators in a forward-looking context. For a general reference on boiler and steam generator technology, see Breeze or Miller.[29,30]

3.1.1   Thermodynamic Properties.   Thermal power generation processes are modelled as ideal Rankine heat engine cycles. Boilers provide the heat addition to the working fluid, which is then expanded through the turbine and cooled in the condenser. This process converts boiler heat into work and low-grade heat. Many thermal power cycles also include at least one reheat process to extract further work. Figure 5 shows a

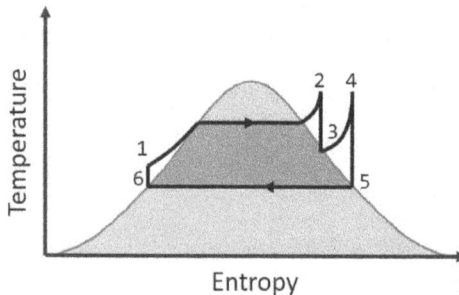

Figure 5   Temperature – entropy diagram for the ideal Rankine cycle with reheat.

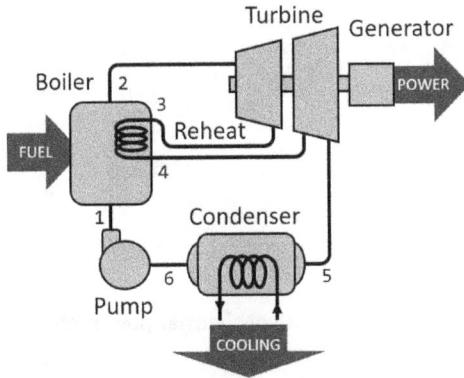

Figure 6 Equipment layout for thermal power generation with reheat.

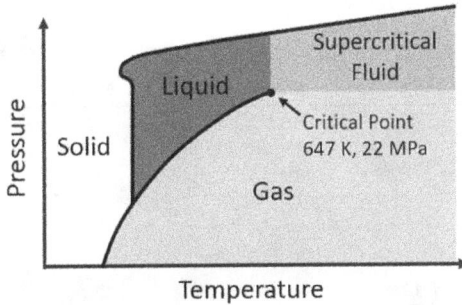

Figure 7 Water phase diagram.

temperature – entropy diagram for an ideal Rankine cycle. Figure 6 shows an equipment schematic of a thermal power generation process with reheat.

Supercritical steam generators generate steam at pressures beyond the fluid's critical point. The critical point of a fluid is the temperature- – pressure extreme of the fluid's phase diagram, shown in Figure 7. Heat addition at pressures beyond the critical pressure causes the fluid to become a supercritical fluid, *i.e.* a phase of matter that displays the properties of both a gas and a liquid. Higher boiler pressures allow more of the added heat to be extracted as useful work rather than exhausted as heat, thus raising power station efficiency. Figure 8 shows the Rankine cycle for supercritical thermal power generation.

Supercritical steam generators are classified according to their range of temperatures and pressures. Generally, the higher the maximum pressure, the higher the thermal efficiency and the lower the emissions. Table 2 shows typical boiler temperatures, pressures, thermal efficiencies and emission rates for subcritical, supercritical, ultra-supercritical and advanced ultra-supercritical boilers.[31–34] Advanced ultra-supercritical steam generators are

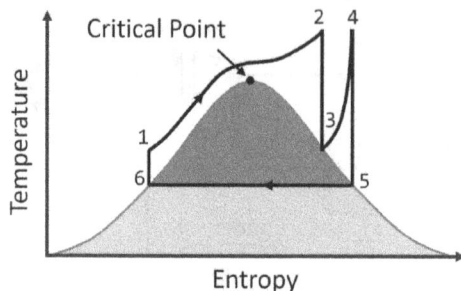

Figure 8   Rankine cycle for supercritical thermal power generation with reheat.

Table 2   Operating and emissions information for coal combustion technologies. (Data drawn from MIT,[31] Riordan,[32] European Union,[33] and Phillips and Wheeldon.[34])

Equipment type	Equipment pressure [MPA]	Equipment temperature [°C]	Thermal efficiency	Emissions				
				CO_2 t MWh^{-1}	SO_x g MWh^{-1}	NO_x g MWh^{-1}	PM_{10} g MWh^{-1}	Hg mg MWh^{-1}
Subcritical	<22	530	<35%	900	272	228	68	39
Supercritical	22–25	540–580	35–40%	851	238[a]	199[a]	68	34[a]
Ultra-supercritical	22–25	580–620	40–45%	836	216	180	64	30
Advanced ultra-supercritical	25–35	700–760	45–52%	763	109	109	36	23[a]
Integrated Gasification Combined Cycle (IGCC)	4.2	1340	38–45%	932	272	228	40	39

[a]Estimated.

the most recent iteration of designs capable of achieving thermal efficiencies in excess of 50% and steam temperatures over 700 °C.

The development of higher-efficiency steam cycles had been limited by availability of materials that were able to withstand the high temperatures of supercritical pressures. Modern chrome- and nickel-based alloys allow all steam generator, turbine and piping systems to withstand the high temperatures for prolonged periods. Welding techniques must also be adjusted to avoid creep and fatigue failures of components made with advanced alloys.

3.1.2   Equipment Layout.   Steam boilers and generators follow typical layouts. Pulverised coal is mixed with preheated combustion air and blown into the centre of a large furnace where it maintains a steady combusting fireball. The fireball transfers heat to boiler water fed through the water wall. Boiling water is collected in the steam drum, which separates water from steam. Boiler designs differ in whether boiler water is recycled

1 Fireball
2 Waterwall
3 Steam Drum
4 Superheater
5 Reheater
6 Economiser
7 Air Preheater
8 Draft Fans

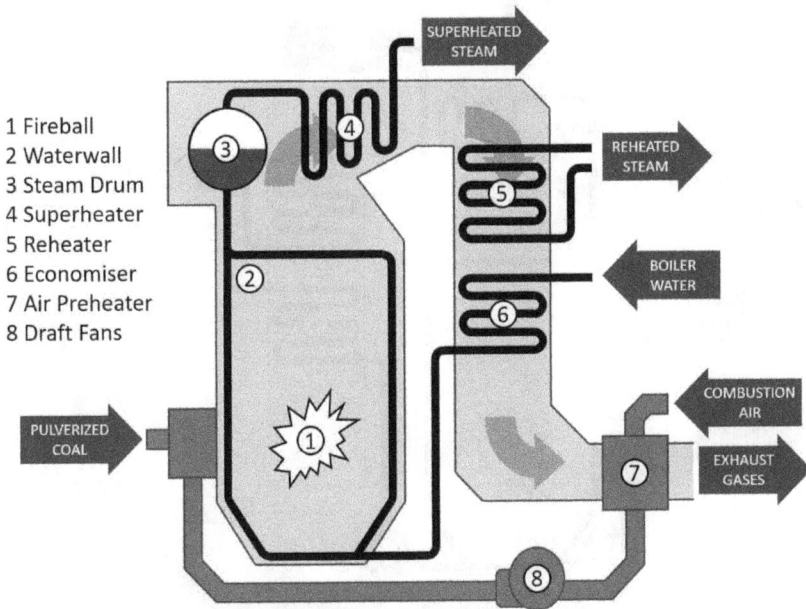

Figure 9 Typical equipment layout for a subcritical steam boiler.

through the water wall and whether water wall circulation is driven by natural circulation or by a feedwater pump. Supercritical steam generators do not have a steam drum, as water does not boil. Figures 9 and 10 show typical equipment layouts for subcritical boiler and supercritical steam generator, respectively.

Superheaters add additional heat to steam from hot combustion gases. Most boiler designs include at least one reheater, which allows lower-pressure turbines to extract further work from steam at higher entropies. Remaining heat is then used to preheat boiler feedwater in the economiser and to preheat combustion air. Cool combustion gases are finally treated for a range of conventional air pollutants and then released to the environment through the stack.

3.1.3 Current State of Deployment. Originally called Benson boilers after their inventor Mark Benson, Siemens AG obtained the patent for super-critical steam generators in the 1920s and began building them in Europe.[35] The high-pressure boilers were able to take advantage of advances in high-pressure turbines and had a basic once-through layout. Supercritical and ultra-supercritical steam generators are now found worldwide.

Supercritical and ultra-supercritical steam generators have become of in-creased interest as policy makers have demonstrated their intention to mitigate greenhouse gas and conventional air pollutant emissions from power generation. The increased efficiency of the steam generation provides utility companies with a technical option for reducing fleet emissions and

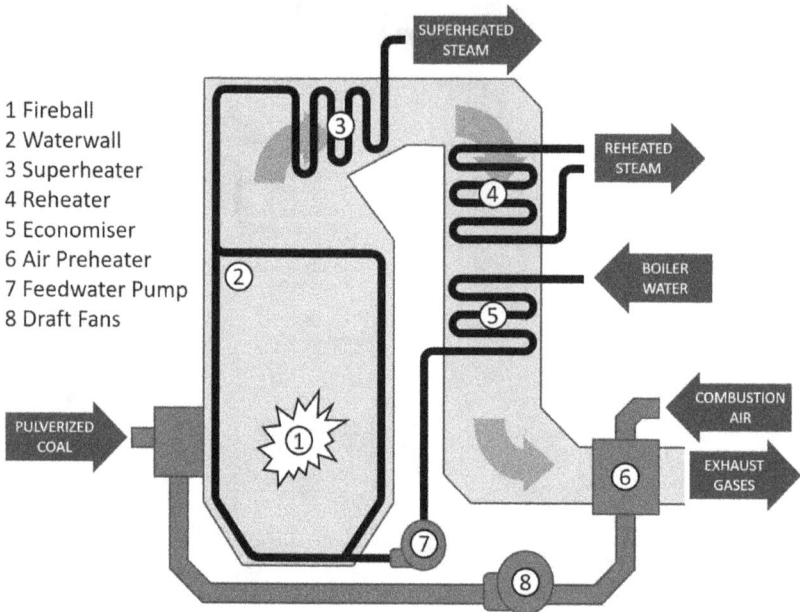

Figure 10 Typical equipment layout for a supercritical steam generator.

policy makers an option for the continued use of inexpensive and/or indigenous coal resources. As a result, supercritical and ultra-supercritical coal-fired power stations in construction and planning pipelines exceed the historic proportion of these technologies in almost every region worldwide (see Figure 11).[36]

3.2 Integrated Gasifier Combined-Cycle

An integrated gasification combined-cycle (IGCC) power plant combines the advantages of a coal feedstock with operating advantages of a combined-cycle gas turbine (CCGT) *via* a coal gasifier. The full energy conversion process includes coal treatment, gasification, gas treatment and gas utilisation *via* both a gas and a steam turbine. IGCCs are required to take advantage of advanced pre-combustion carbon capture technologies.

The main subsystems of an IGCC plant are the gasification plant, raw-gas cooling (*via* water quench or heat recovery systems), water gas shift reactor, gas purification system (with sulfur removal/recover and potentially carbon dioxide removal), an optional air separation unit and combined-cycle unit (with a gas turboset, heat recovery system generator and steam turboset). The commercially operational IGCC plants all employ slightly different gasification technologies; gas cooling and clean-up systems; and processes to integrate steam. A particular difference is the integration of the gas turbine with the air separation unit (ASU). The air separation unit supplies the

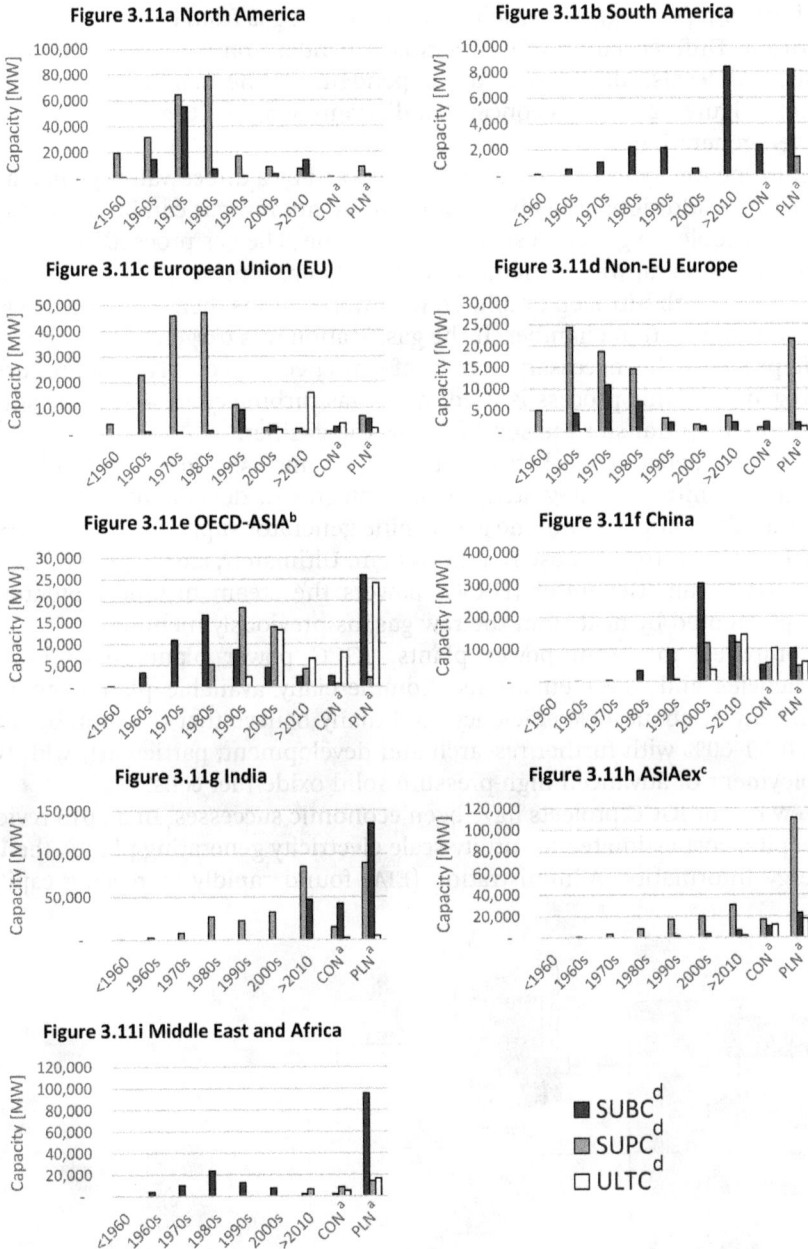

Figure 3.11a North America

Figure 3.11b South America

Figure 3.11c European Union (EU)

Figure 3.11d Non-EU Europe

Figure 3.11e OECD-ASIA[b]

Figure 3.11f China

Figure 3.11g India

Figure 3.11h ASIAex[c]

Figure 3.11i Middle East and Africa

■ SUBC[d]
■ SUPC[d]
□ ULTC[d]

a) CON – under construction; PLN – in planning
b) OECD-ASIA – Japan, South Korea, Australia, New Zealand
c) ASIAex – Non-OECD Asia excluding India and China
d) SUBC – Subcritical boiler; SUPC – Supercritical steam generator; ULTC – Ultra supercritical steam generator

Figure 11 Global deployment and pipeline of coal-fired power stations by region. (Data drawn from Oxford Smith School[36] with permission,)

gasifier with pure oxygen, which prevents the syngas from being diluted with nitrogen. Differentiation in gasifiers is dependent on cost, efficiencies and emissions levels, all of which are dependent on the fuel stock and application. Figure 12 shows a conceptual diagram of an IGCC equipped with an oxygen-generating ASU.

After the raw syngas is produced, it is cooled by a direct water quench and sometimes additional sensible heat from the hot raw gas can be captured in a syngas cooler to generate steam for a turbine. The gas proceeds to the gas treatment system, where chemical pollutants are removed. The clean gas is next mixed with nitrogen or diluted with water and is then supplied to a gas turbine combustion chamber. If the gasification was oxygen-blown, the ASU unit provides the necessary supply of enriched oxygen. The co-produced nitrogen from this process is used in the gas turbine cycle, and some small amounts help transfer the solid fuels to the gasifier.

The interdependency between the IGCC and ASU can be called air-integrated, nitrogen-integrated, or non-integrated, dependent on how and whether the compressor of the gas turbine generator supplies air for the ASU and how the nitrogen is used in the system. Ultimately, steam from the Heat Recovery Steam Generator (HRSG) powers the steam turbine, potentially complemented by heat from the raw gas, as previously mentioned.

Compared to steam power plants, IGCC power plants have higher efficiencies and lower emissions. Commercially available plants produce electricity at about 40% efficiency, and many believe that this number can reach 50–60% with further research and development, particularly with the deployment of advanced high-pressure solid oxide fuel cells.

Few recent IGCC projects have been economic successes. In a 2013 review of capital cost estimated for utility-scale electricity generating plants, the US Energy Information Administration (EIA) found rapidly increasing capital

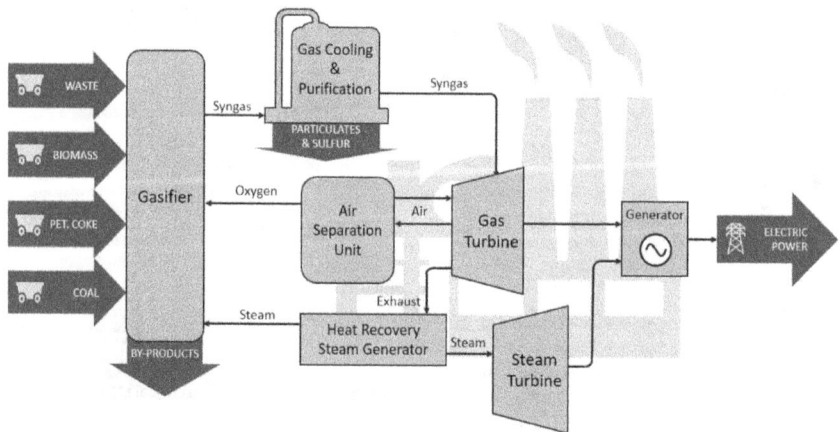

Figure 12　Conceptual flow diagram of an Integrated Gasification Combined Cycle (IGCC) power station with integrated Air Separation Unit (ASU).

Table 3 Global deployment of Integrated Gasification Combined Cycle (IGCC) power plants. (Data drawn from the National Energy Technology Laboratory.[37])

Country	Number of projects
United States	8
United Kingdom	4
India	3
China	2
Japan	2
Botswana	1
German	1
Pakistan	1
Saudi Arabia	1
South Korea	1

costs of almost 20% for IGCC plants, with or without Carbon Capture and Storage (CCS). High production costs growing from ballooning capital costs make the cost of electricity generated by IGCC power plants unable to compete with conventional coal power plants, as well as renewables and natural gas. Due to their complex structure, they face much higher investment costs, and thus higher production costs—of course, the higher efficiency may lead to total average costs for unit of power generation to be lower.

Table 3 shows the global deployment of IGCC power plants according to the National Energy Technologies Laboratory (NETL) database.[37]

Between the US and UK's 12 listed projects, only one is actually active and 8 have been substantially delayed or cancelled. The most famous project in the US, Kemper, has consistently faced delays and cost overruns. The largest growth of IGCC power plants will likely come from Japan, with a pipeline of over 40 projects to balance the decline of its nuclear generation capacity; many of these are expected to be IGCC plants. Optimism in the sector is built on research of growing efficiencies, which will likely be driven by greater optimisation of the integration of gasification and power generation, advances in gas-turbine technologies (such as higher pressure ratios on turbine entry), further development of the cleaning process, and advanced air separation technologies.[38]

3.3 Solid Oxidisers and Chemical Looping

Chemical-Looping Combustion (CLC) is a cycling process where an oxygen carrier transfers outside oxygen to a fuel *via* oxidation–reduction reactions. This topic is covered extensively by Anthony in Chapter 7 of this volume.[9] CLC aims to completely oxidise fuel to carbon dioxide and steam while producing heat for power production. It has a variety of uses, including generating electricity, hydrogen, chemicals and liquid fuels.[39] This process includes interconnected fuel and air reactors (a reducer and an oxidiser),

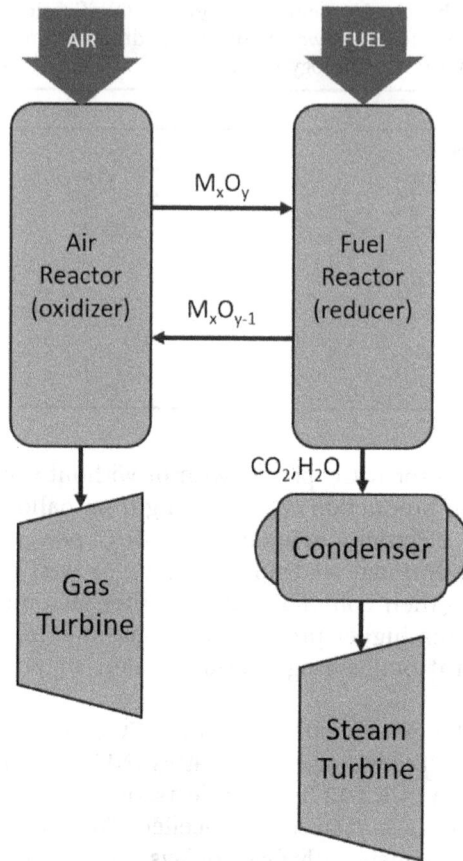

Figure 13　Conceptual diagram of chemical-looping combustion.

where a solid oxygen-carrier is reduced and then re-oxidised in the fuel and air reactor, respectively.[40] The power generation process in a CLC-based power plant is as follows, and the process is set out in Figure 13.

There are two main processes for solid fuel chemical looping: syngas chemical looping combustion/gasification and direct solid fuel combustion/gasification. In the former, coal is initially gasified to syngas; the later feeds the coal directly the fuel reactor.

Chemical Looping is considered a promising CO_2 capture technology. A commercial scale 550 MW coal direct-looping power plant has demonstrated 96.5% CO_2 capture with a 28.8% increase in electricity costs.[41] Chemical looping can remove nitrogen from the process, reducing NO_x emissions and producing a flue gas stream of pure CO_2 and H_2O. That is, the CLC process performs like an oxy-fuel combustion process but with lower cost as it removes the need for an independent energy-intensive air separation unit. A CLC process can be integrated with an IGCC process to deliver pure hydrogen to a gas turbine power cycle (see Figure 14).

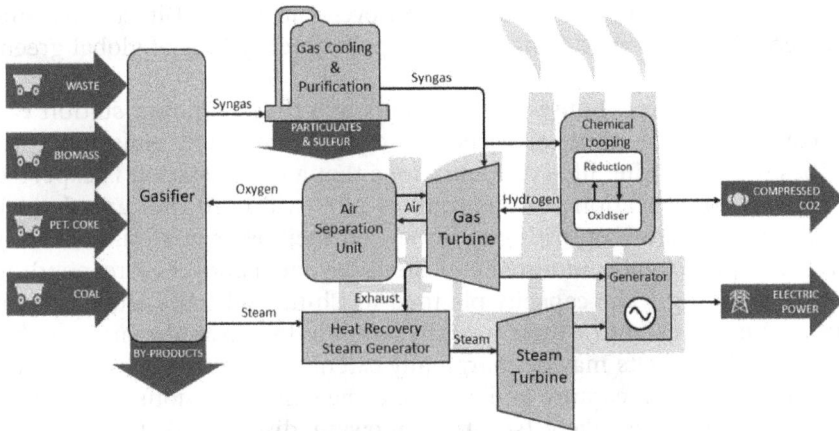

Figure 14 Conceptual diagram of an Integrated Gasification Combined Cycle (IGCC) power station with integrated chemical-looping combustion.

4 Coal-fired Power Stations in the 21st Century

The future of coal-fired power is intrinsically linked with the future of the environment and society's responses to environmental change. This section will explore a series of environment-related challenges facing coal-fired power generation in the 21st century. The committed greenhouse gas emissions of coal-fired power stations are low-hanging fruit for reduction *via* policy interventions. The water consumption of power stations competes with other industries, agriculture and domestic use. Conventional air and water pollutants from coal-fired power stations threaten human and environmental health. Climate stresses in the form of heat waves and water shortages threaten power station operation. Renewables are displacing coal-fired power stations in the merit order curve, and coal-fired power is insufficiently flexible to partner with renewables to alleviate intermittency. Operating and capital expenses of coal-fired power are particularly sensitive to carbon pricing and CCS inclusion. Few coal-fired power stations are retrofittable with CCS, and the technology is making little progress towards being attractive as a climate mitigation option. Development and private finance is becoming less available to the builders of coal-fired power stations.

4.1 Committed Carbon Emissions and Stranded Assets

The convergence of international climate policy around a 2 °C limit to average warming and scientific consensus around the transient climate response to cumulative greenhouse gas emissions has enabled the establishment of a notional carbon budget. These issues are considered extensively by Kimmel and Cleetus in Chapter 5 of this book.[42] Reserves of fossil fuels far exceed this carbon budget,[43] raising concerns of stranded fossil-fuel assets and a financial bubble attributed to their over-valuation.[44] Direct emissions from coal-fired power stations produce approximately 31% of global greenhouse gas emissions.

The life-cycle greenhouse gas emissions of a thermal power station can be compared to a notional carbon budget. 'Committed emissions' describes the remaining life-cycle emissions that a power station is expected to emit. A power station built as soon as 2017 will fulfil the capital stock requirements that, if operating to or beyond their average lifetime of forty years, will push total committed emissions beyond a budget corresponding to 2 °C warming.[6] The capacity planned in China and India alone may, if run to the full extent of their useful economic life, exceed emissions beyond 2 °C.[45] Budgets may be marginally extended by either significant investment in carbon capture and storage or negative emissions technology, neither of which is likely (see the following discussion). Constraining warming to 2 °C may cause some power stations already in operation, and certainly many of those in planning or under construction, to become stranded assets.

Stranded assets are assets which have suffered from unanticipated or premature write-downs, devaluations, or conversion to liabilities.[46] The creditors and corporate owners of these power stations are exposed to reductions of cash flows or profitability under changing market conditions due to, among other factors, taxes or quotas on carbon emissions, competition from lower-carbon generating technologies, and changing consumer appetites for high-carbon power.[22,47,48] Financial returns from affected power stations may fail to appropriately compensate risk-bearing investors or may even fall short of the recovery of investment costs. The definition of environmental risk faced by investors is now expanding to include a wide range of environment-related concerns, from the acute and chronic physical impacts of climate change, to the transition risks embedded in market, technology, policy, legal and reputational responses to environmental change. The work of the Bank of England,[49] the Task Force for Climate-Related Financial Disclosures,[50] and many others, are driving a new sophistication in the disclosure of and responses to environmental risk.

4.2   Water Consumption

Coal-fired power generation places significant pressure on local water supplies, as coal-fired power stations are second only to nuclear power stations in water use. Water is primarily used for cooling, as a heat sink for the condensation of the turbine working fluid; however, power stations also use water for coal pre-treatment and pollution control. The largest factor in determining the water-efficiency of stations is the type of cooling system installed. Secondary factors are the ambient temperature and station efficiency. Table 4 shows water use in various types of power generation based on cooling technology.[51]

Table 4 Water use in by various power generating options. (Partially reproduced from the Electric Power Research Institute with permission.[51] © 2008 Electric Power Research Institute, Inc. All rights reserved.)

| Fuel-type | Once-through | Cooling technology | | Dry cooling |
| | | Closed-cycle (wet) | Hybrid (wet/dry) | |
		(gallons per MWh)		
Coal	95 000–171 000	2090–3040	1045–2755	~0
Gas	76 000–133 000	1900–2660	950–2470	~0
Oil	76 000–133 000	1900–2660	950–2470	~0
Nuclear	133 000–190 000	2850–3420	N/A	N/A

Cooling technologies include once-through cooling, and wet, hybrid and dry cooling towers. Once-through cooling draws water from surface or groundwater, uses it to absorb waste heat, and then returns heated water to the environment. Wet, hybrid and dry cooling towers transfer waste heat to airflows which are returned to the atmosphere. Wet cooling towers inject warm water into these air flows, transferring both sensible and latent heat to the air with the requirement of additional make-up water. Dry cooling towers transfer only sensible heat to cooling air, with flow separation – for example, by a heat exchanger. Hybrid cooling towers combine aspects of both dry and wet cooling towers.

Inefficiencies in thermal power stations cause increases in the amount of waste heat required to be ultimately transferred to the environment. Power station water consumption is therefore sensitive to the station's efficiency. Several climate-related factors exacerbate water consumption, *via* impairments of plant efficiency, such as the increase in ambient temperature and the implementation of CCS. Climate stresses are discussed further below.

CCS also has a number of non-efficiency impacts on power station water consumption. Both pre- and post-combustion technologies require the recycling of CO_2 absorbent material *via* temperature manipulation. Pre-combustion CO_2 capture also requires additional water for water shift reactions. Even oxy-fuel combustion may require an additional water footprint in the heat transfer processes of the cryogenic preparation of the oxygen feedstock. Figure 15 shows the difference in water consumption between subcritical, supercritical, and IGCC power stations with and without CCS, as well as nuclear, gas, and renewable generating options.[52]

The availability of water resources is a key concern for coal-fired power stations in the 21st century. Climate change is fundamentally disrupting the acute and chronic availability of water for thermal power generation. Environmental change also causes a wide range of covariant affects in other human activities which similarly rely on water, particularly agriculture and industry, and domestic use subject to population changes and migration. Water is often included in a 'nexus' of system impacts in food, water and energy provision.[53] Finally, water policy has endemically failed to price or regulate water extraction to avoid the 'tragedy of the commons' nature of shared water resources (*e.g.* water policy review).[54]

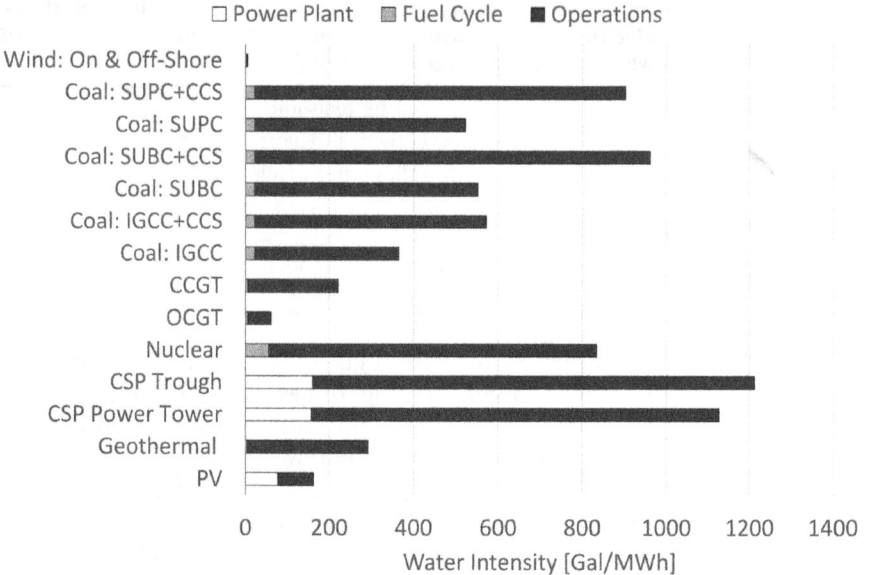

Figure 15   Life cycle water consumption of generating options (Supercritical, SUPC; Subcritical, SUBC; Carbon Capture and Storage, CCS; Integrated Gasification Combined-Cycle, IGCC; Combined-Cycle Gas Turbine, CCGT; Open-Cycle Gas Turbine, OCGT; Concentrated Solar Power, CSP; Photo Voltaics, PV).
(Data drawn from J. Meldrum *et al.*[52])

Previous research shows that there is strong evidence to suggest that unavailability of water resources is a legitimate concern to the profitability of power stations.[51] In India, coal-water risks have forced nationwide blackouts and water shortages that restrict plants from operating at full capacity and have been shown to quickly erode the profitability of Indian power stations.[55] In China, attempts to abate local air pollution in eastern provinces have pushed coal-fired power generation into western provinces, where there is extreme water scarcity and shortages are expected.[56] Water availability in the 21st century will drive decision making in the technical design of coal-fired power stations and their cooling technologies; conflicting priorities over water use for electricity generation, agriculture, industry and domestic consumption will drive political economy tensions within and between countries.

4.3   Pollution Formation

A wide range of conventional pollutants dangerous to human and environmental health are emitted by coal-fired power stations. In the 21st century, the attention given to these emissions by policy makers and the concerned public will continue to grow. Growing populations more concentrated in urban centres will demand toxin-free air to breathe and water to drink. The increasing availability of technological alternatives, like lower-emissions gas-fired power or emissions-free renewable power, reduces zero-sum trade-offs between the availability of electricity and human health. This topic is covered extensively by Gottlieb and Lockwood in Chapter 4 of this volume,[57] so is considered only briefly here. See Vallero for a general reference on air pollution and air pollution control.[58] See Zhang for a general reference on coal combustion wastes.[59]

4.3.1 Air Pollutants. The majority of pollution emitted by a coal-fired power station is air pollution. The combustion of coal creates a wide range of pollutants, notably particulate matter, volatile organic compounds, oxides of nitrogen and sulfur, and mercury. Pollution control technology can reduce the concentration of these compounds in the final exhaust released to the environment. These pollutants are significantly detrimental to human and environmental health. Their airborne nature allows them to both impact large areas or to be concentrated by weather patterns and geography in small areas.

Volatile organic compounds (VOCs) are airborne organic compounds produced as a result of incomplete combustion. VOC is a general term encompassing the various gaseous derivatives of aldehydes, ketones, aliphatic compounds, benzene and polycyclic aromatic hydrocarbons (PAHs). VOCs have a range of respiratory and carcinogenic affects, and react with nitrogen oxides to form ground-level ozone, a component of smog. VOC concentrations can be controlled by passing flue gases through an afterburner, which completes the combustion of the remaining organic compounds.

Nitrogen oxides include nitric oxide (NO), nitrogen dioxide (NO_2), and nitrous oxide (N_2O), referred to collectively as NO_x. Some amount of NO_x is formed in any combustion process where air is the oxidiser – the heat of combustion causing otherwise inert nitrogen to oxidise. The larger part of NO_x in coal flue gas is derived from nitrogen compounds in coal. NO_x reacts with airborne water to form nitric acid, leading to acid rain. NO_x also reacts with VOCs to form ground-level ozone, a component of smog. NO_x can be removed from flue gas using a catalytic converter.

Sulfur oxides (SO_x), particularly sulfur dioxide (SO_2), is caused by the oxidisation of sulfur in fuels. Many coal fuels are sulfur-rich leading to the substantial formation of SO_x. SO_x reacts with airborne water to form sulfuric acid and acid rain. SO_x can be removed from flue gases by sorbent injection, dry flue gas desulfurisation, and wet flue gas desulfurisation.

Particulate matter (PM) formed in coal combustion is a collective name for small airborne liquid and solid particles mixed in otherwise gaseous exhaust gases. In exhaust gases from coal combustion, particulate matter is composed of inorganic ash particles, uncombusted solid carbon (soot), and droplets of liquid-phase NO_x and SO_x. PM is categorised into particle size, generally into PM_{10} and $PM_{2.5}$, referring respectively to particles with diameters less than 10 micrometers and 2.5 micrometers. The main health impact of PM is its interference with lung function. The smaller the PM, the further the PM is able to penetrate the lungs and the greater the health impacts. PM also bears a health risk according to its constituent parts, including toxicity and carcinogenic effects. PM can be removed from flue gases by treatment in a baghouse, which passes dirty gas through fabric filters, or by an electrostatic precipitator which removes airborne particles using an electrostatic charge.

Mercury is naturally occurring in coal. With its low boiling point, mercury is readily released into the environment among flue gases. Mercury derivatives are toxic and can interfere with reproductive, cardiovascular and neurological health. Fat-soluble organic mercury compounds are persistent in organisms and thus bioamplify along the food chain. This poses a particular threat to humans with fish-based diets. Mercury removal is a side benefit of the other pollution control technologies discussed previously. Chemical additives like calcium bromide and activated carbon can be added to flue gases to further capture mercury prior to release.

4.3.2 Land and Water Pollutants. Coal-fired power stations emit a number of land and water pollutants that affect the health of local environments and populations. Land and water pollution result largely from the improper containment of toxic materials related to coal combustion, such as waste coal, ash and scrubber sludge, and gypsum ponds. Land and water pollution is a much larger concern at coal mining and preparation sites; however, many coal-fired power stations are co-located with these other operations.

Capture of conventional air pollutants reduces them to a solid waste which is storable in landfill. Coal Combustion Waste (CCW) includes captured ash and sulfur slurries from exhaust gas treatment, and slag and ash residues from coal combustion. It is the second largest solid waste stream in the US after municipal solid waste and contains concentrated amounts of mercury, arsenic, beryllium, cadmium, lead, sulfates and sulfides. These compounds are toxic to human and environmental health. Spillage, improper storage, and seepage can cause the release of these toxins into the land, surface water and groundwater.

4.3.3 Pollution Formation in the 21st Century. Air pollution has emerged as a primary concern of activists and policy makers in the 21st century. Urbanisation trends have amplified exposure to and awareness of the health impacts of air pollution from coal-fired power stations. Many of these trends occur in countries such as China and India where coal-fired power is expected to support rapid urbanisation and industrialisation. Policy makers face the choice to intercede with operating and planned coal-fired power stations in order to reduce the impact of pollution on their populations.

Air quality reduces average life expectancy in China by 25 months. Recognising the urgency of this public health crisis, China enacted the Action Plan for Air Pollution Prevention and Control (APPC) in 2013. Measures enacted under the APPC include the ban of high-ash and high-sulfur coal imports, expanded coal washing, and coal consumption caps in three heavily populated regions. Coal-fired power stations must be equipped with pollution controls and will be banned from operating in urban and suburban areas from 2020 onwards.

Average life expectancy in India is reduced by 23 months by air pollution. $PM_{2.5}$ concentrations in India have been steeply increasing since 2010 while those in China have stabilised.[60] Policy makers in India have not taken as extreme steps to curb the increase in pollution with the increase in fossil energy consumption. Crisis-level pollution in New Delhi in November 2016 caused the closure of schools; the closure of local power stations and industrial sites; suspension of construction activities; the deployment of emergency water spraying services; and mass public protests.[61]

In the 21st century, air pollution control will continue to increase as a priority for policy makers. Some policy makers may also use controls on conventional air pollution to encourage the early retirement of carbon-intensive power stations. In the US, controls on conventional air pollutants were forcing upgrade-or-retire decisions in coal plant operators well ahead of the fall in gas prices which would seal the fate of US coal-fired power stations.[62]

4.4   Climate Stresses

There is a significant negative feedback loop between climate change caused by greenhouse gas emissions of thermal power generation (among other anthropogenic sources) and the efficiency and profitability of thermal power generation. A growing number of event studies in climate science are able to appropriately attribute the contribution of anthropogenic climate change to individual extreme weather events.[63] These weather events—including droughts, floods and heat waves—are detrimental to the operating of thermal power stations and those attributable uniquely to climate change can be quantified.[64,65] Examples include event studies of droughts in Africa that were able to demonstrate both natural and anthropogenic contributions to weather variability,[66,67] and extreme event attribution of heat-related events in East and Southern Africa, Europe, South-east Asia and Oceania.[68,69]

The impacts of droughts on thermal generation facilities and power systems as a whole have been well documented. Thermal generation plants have a high demand for water across cooling, fuel processing and emissions control. A 2012 US Department of Energy analysis found that there were 143 GW of thermal generation at high-risk locations exposed to drought-related shutdowns.[70] A major benefit of the shift of the US fleet toward natural gas and renewables has been the decrease in the vulnerability of the US electricity system to water scarcity and higher water temperatures.[71]

Thermoelectric power dominates water withdrawals in the US and even moderate droughts can have severe economic damages in the electricity sector—a moderate drought in the US could cause economic damages of $51 million.[72,73]

Recently, water shortages have caused shutdowns across India, where over 170 GW of coal power plants (40% of the India fleet) are in areas of high water stress due to increased droughts and variations in annual rainfall.[74] In 2016 alone, water-driven shut-downs in India drove a loss of over 8.7 billion KWh of generation, or 1% of total generation in the country over the year, concentrated in specific areas, and oftentimes areas where additional plants are being developed.[75] Similar analysis in China and the US found that 45% and 6.85% or their respective coal fleets to be in areas of high levels of water stress or in areas that are already suffering significant over-withdrawal.[76] Competition for scarce water resources between thermal generation plants and other uses, such as agricultural productivity, has been well documented in South Africa as well, including rulings that have suspended the development of the 1.2 GW Thabametsi coal-fired station and growing challenges to the water use of the 4.8 GW Medupi station.[77] Worldwide, decline in surface water availability and increase in water temperatures are expected to reduce thermoelectric power station capacity by 5% in the 2020s and 12% by the 2050s.[78] Fuel switching to gas or renewables will be a key way to mitigate generation capacity vulnerability.[79]

Heat stress induced by climate change is already impacting thermal power generation. Both acute (short-term) and chronic (long-term) temperature increases can be detrimental to power station operation. Short-term heat waves can raise air and water temperatures, which reduces oxygen availability and cooling efficiency and can lead to the violation of discharge regulations. Heat waves also increase air conditioning loads; in many countries peak grid load occurs during hot days in the summer. Heat stress also impacts the efficiency of transmission and distribution infrastructure as temperature increases wire resistivity and decreases transformer efficiency. These coincident phenomena can cause crisis for generators and grid operators.

There are numerous examples of the impact of heat waves of thermal generation and grid infrastructure.[80] In 2006, heat waves in California resulted in the failure of over 2000 transformers, causing shortages affecting 1.2 M people and leaving 80 000 without power for days. High intake water temperatures during a 2007 heat wave caused multi-day shutdowns of thermal generating stations in the Southeast United States. In 2009, a heat wave in Australia caused the shutdown of a regional interconnector and blackouts lasting up to 2 days for 500 000 Melbourne residents.

Chronic heat stress fundamentally reduces the operating efficiency of thermal power stations. Even under ideal operating conditions, the higher the ambient temperature, the less efficient the power station will be. Power output is impaired by between 0.15% and 0.5% *per* 1 °C increase.[81] On a 'business as usual' warming pathway, 1 °C of warming is projected to occur every 25 years, making these losses a material concern for plant owners and investors.

4.5 Generating Flexibility and Dispatch Merit Order

In most electricity markets, power is dispatched by 'merit order' – the ordered ranking, from lowest to highest, of the bid prices of the generators. These prices are broadly reflective of the marginal costs of generation for the plants. Nuclear and renewable generating options have almost no marginal costs and so can bid prices that are low (or even negative), knowing that incumbent generators must bid higher to cover their costs – giving them generating priority and happily accepting an equilibrium price well above their bid price. This is illustrated in Figure 16.

As renewable power development continues to grow, renewable generating options are added to the lowest marginal cost stages of the merit order curve and reduce equilibrium power prices by pushing incumbent generators to the right. This is illustrated in Figure 17. As a first-order effect, reduced equilibrium prices decreases margins for both incumbent and emerging renewable generators.

Furthermore, as renewable generators displace conventional thermal power, the overall capacity factors of thermal generation plants decrease. The levelised cost of electricity (LCOE) of incumbent plants increases as it becomes more weighted by fixed rather than variable costs. This effect is more detrimental to plants with high capital costs, like thermal generating plants equipped with CCS, as shown in Figure 18.[82–84]

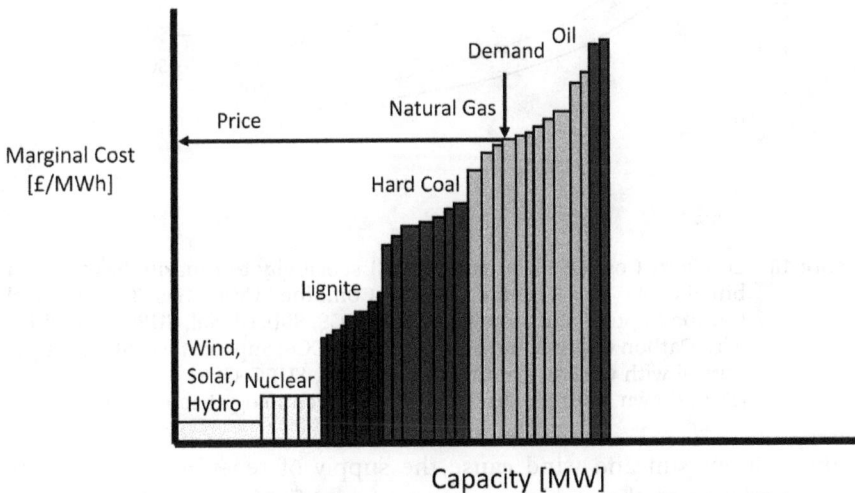

Figure 16 Merit order of generating options (MC = Marginal Costs).

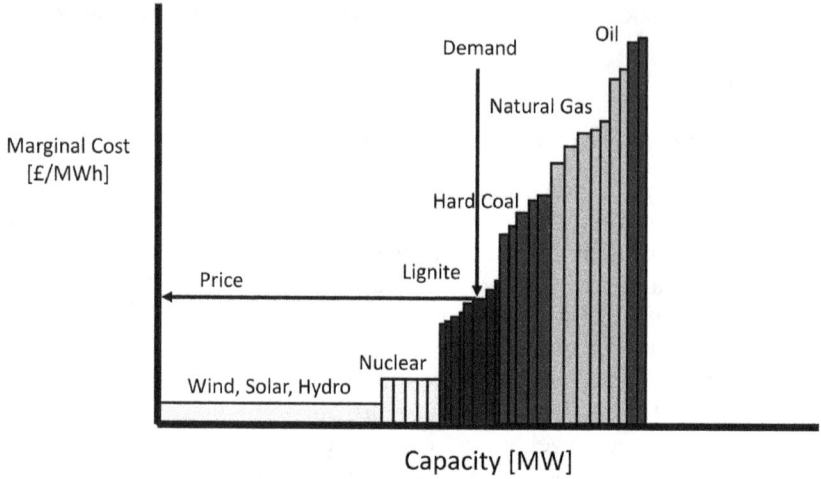

Figure 17 Merit order of generating options with enhanced deployment of renewables.

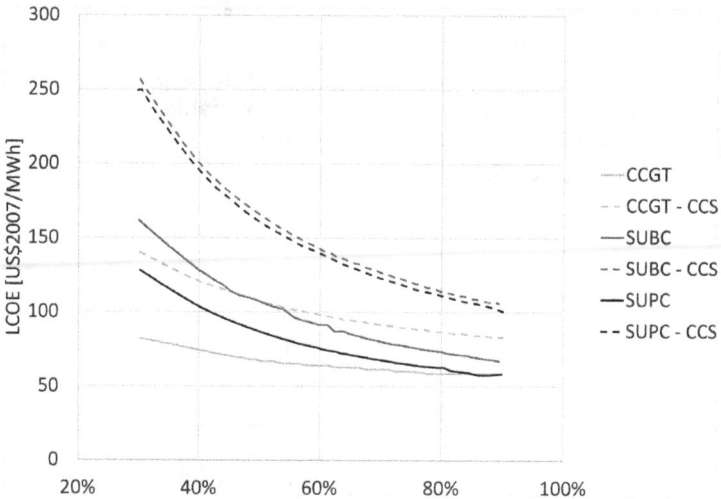

Figure 18 Levelised Cost Of Electricity (LCOE) sensitivity to capacity factor (Combined-Cycle Gas Turbine, CCGT; Combined-Cycle Gas Turbine with Carbon Capture and Storage, CCGT – CCS; Subcriticial, SUBC; Subcritical with Carbon Capture and Storage, SUBC–CCS; Supercritical, SUPC; Supercritical with Carbon Capture and Storage, SUPC–CCS).
(Data drawn from the National Energy Technology Laboratory.[82–84])

Intermittent sun and wind cause the supply of renewable power to be volatile relative to the stable operation of coal-fired power stations. This volatility in supply causes the equilibrium price and marginally dispatched generating unit to change over the course of a day. Power stations are increasingly called to ramp up and ramp down power output.[85] The onset of renewables has increased the incentive of generators to be flexible in their ability to supply power even while decreasing equilibrium prices. Governments are looking for policy options to incentivise flexibility in electricity markets, including flexible generation technologies, demand response, interconnectors and electricity storage.[86] New price mechanisms are being introduced to better match electricity demand and supply, allowing generators, large consumers, and even homes and appliances to respond to electricity pricing in real time.

It is uncertain whether coal-fired generators will be able to meet the flexibility demanded by the intermittency of renewables. Fossil fuel thermal generation is more dispatchable than nuclear power in balancing the intermittency of renewables; however, coal may not be the fossil fuel of choice to capitalise on the new incentives of flexibility. This role is easily filled by combined-cycle gas turbines (CCGTs) and open-cycle gas turbines (OCGTs), despite gas being a more costly generating option in most of the world.

The speed with which a generator can sync to the grid and begin producing useful power is called its 'ramp rate'. Ramp rates from different start points are shown in Table 5.[87] A 'hot' start might be defined when restart occurs after only a short period of inactivity (<8 h), 'warm' (8–48 h), and 'cold' (>48). Across technologies, the time from notice to sync for coal is significantly higher than gas.

The cycling of fossil generation plants – the operation of generation units at varying load levels in response to system load requirements – is expensive and is a primary cause of rising operating and maintenance (O&M) costs in conventional generators.[88] Frequent heating and cooling of power station equipment causes thermal fatigue, which requires plant downtime and capital expense to fix. Thermal fatigue is also exacerbated by the temperature differential through which the material is being cycled. Which is to say,

Table 5 Ramp rates for various thermal generating options. (Data drawn from Lunn.[87])

Temperature	Technology	Notice to sync (minutes)	Sync to full load (minutes)
Hot start (<8 h)	Coal	80–90	50–100
	Existing gas CCGT	15	40–80
	Modern gas CCGT	15	25
	Large gas OCGT	2–5	15–30
Warm start (8–48 h)	Coal	300	85+
	Gas CCGT	15	80+
	Large gas OCGT	2–5	15–30
Cold start (>48 h)	Coal	360–420	80–250
	Gas CCGT	15	190–240
	Large gas OCGT	2–5	15–30

under the same cyclical loading conditions, supercritical steam generators operating at higher temperatures and costs will suffer proportionately increased costs. Coal-fired power stations may, therefore, not be designed to be both efficient *and* flexible. Emissions constraints can also make load-following difficult for generators – the cycling process may cause much higher than average emissions for short periods.

4.6   Operating and Capital Expenses

For coal to remain a competitive power generation source in the 21st century, it must be cost-competitive with its many lower-emission substitutes. Coal, being the most carbon-intensive generating option, will be most affected by policy efforts to appropriately price or control carbon emissions. Any decarbonisation pathway-centric policy, whether through taxes or command-and-control regulation, will make it more expensive to develop coal-fired power generation projects. Power projects are long-term cash flow projects, and discounted cash flow analysis used in investment decision-making may undervalue future policy changes. Investment theses may differ from region to region and market to market; however, it is abundantly clear that coal-fired power remains a low-cost power source only if the cost of CCS retrofitting or biomass co-firing, as might be brought on by a future policy change, is resolutely ignored.

In the United States, the Energy Information Administration measured Base Overnight Costs across 2016 for a series of technologies and found the most expensive to be landfill gas (US\$8509 per kW), fuel cells (US\$6252 per MWh), coal with 90% sequestration (US\$5072 per MWh), offshore wind (US\$4648 per MWh), and coal with 30% sequestration (US\$4586 per MWh); overnight costs exclude interest accrued during plant construction and development.[89] For operating expenses, the highest fixed costs were for landfill gas (US\$410.32 per kW), geothermal (US\$117.95 per MWh), biomass (US\$110.34 per MWh), advanced nuclear (US\$99.65 per MWh), and coal with 90% sequestration (US\$80.78 per MWh); for variable costs, the top expenses were with fuel cells (US\$44.91 per MWh), advanced combined cycle turbines (US\$10.63 per MWh), coal with 90% sequestration (US\$9.54 per MWh), and landfill gas (US\$9.14 per MWh). Amongst thermal power plants, combined fixed and variable O&M costs increase from conventional coal power plants to those with IGCC and those with IGCC and CCS.[90]

Some of the most advanced work on the topic has been done by the National Energy Technology Laboratory (NETL). Their research has found rising costs with CO_2 capture across both total overnight costs (TOC) and levelised cost of electricity (LCOE), as seen in Table 6.[91]

These increasing costs are also demonstrated by Stacy and Taylor,[92] who find that existing coal-fired plants generate a LCOE of US\$38.4 per MWh, while new coal plant costs range between US\$80 per MWh and US\$97.7 per MWh due to operational frequency differences, and the authors conclude that existing generation resources are better positioned to deliver low cost

Table 6 Additional costs of Carbon Capture Storage (CCS) for Total Overnight Cost (TOC) and Levelised Cost of Electricity (LCOE). (Data drawn the National Energy Technology Laboratory.[91])

Plant type	TOC (US$ per kW)		LCOE (US$ per MWh)	
	No CCS	CCS	No CCS	CCS
CCGT	718	1497	59	86
SUBC coal	2010	3590	59	108
IGCC	2505	3568	77	112

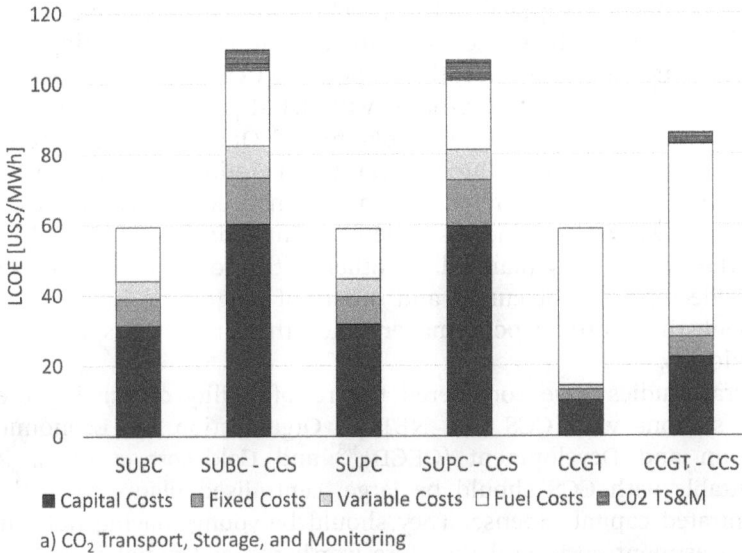

a) CO$_2$ Transport, Storage, and Monitoring

Figure 19 Levelised Cost Of Electricity (LCOE) breakdown for generating options. (Data drawn from the National Energy Technology Laboratory.[93])

electricity, preferring that plants are fixed to extend their life rather than seeing new capacity built.[92] Complimented by the fact that the global share of coal plants older than 20 years is at 60% (and higher in the US and Europe), it is likely more economical for existing generators to be retrofitted and run for an extended life span than for new power plants to be built, especially on established grids. Power generation projects with CCS experience an increase in total energy costs of 46–85% (see Figure 19).[93]

4.7 Carbon Capture and Storage (CCS) Retrofitability

CCS may be able to substantially reduce the emissions of existing or new-build power stations in line with constrained emissions scenarios consistent with climate targets. Anthony in Chapter 7 of this volume provides a detailed examination of CCS and its role in the future of coal in the 21st century,[9] so we will limit our contribution in this section to argumentation. We argue that it is unlikely that CCS will make any substantial contribution to emissions mitigation or carbon budget extension, and that the efforts of policymakers, technologists and financiers are better spent elsewhere.

A substantial portion of new-build coal-fired generation must be equipped with CCS if coal-fired generation is going to remain a substantial portion of a global generating mix compatible with a pathway to 2 °C. Existing power stations would also need to be retrofitted in order to meet a 2 °C warming target.[1] Table 7 shows the amount of carbon dioxide which would need to be sequestered according to publications of the IEA and IPCC by 2040.[7,94,95]

Projections of CCS growth rely on technology learning curves to reduce CCS prices sufficiently to allow CCS to compete with other mitigation options on costs. There are currently 17 operating CCS projects, sequestering a total of 32 Mtpa. A further 5 projects with 3.4 Mtpa are under construction and 16 projects with 26 Mtpa are in planning.[96] Of the operating and under-construction projects, only three are power generation facilities and most of the projects use the sequestered carbon for enhanced oil recovery. Achieving CCS targets will require increased deployment orders of magnitude larger than what is currently planned. As other mitigation options have shown, rapidly decreasing cost curves and orders of magnitude increases in deployment, the efforts of policymakers and other are better spent on these technologies.

Several studies have considered the retrofitability of existing thermal power stations with CCS (see NETL,[97] Organization for Economic Co-operation and Development (OECD),[98] and Caldecott et al.[22]). Plants retrofittable with CCS should be large, centralised plants to justify the concentrated capital expense. They should be young, having not paid off their investment costs and therefore worth the additional investment to avoid becoming stranded. They should be efficient, so that they might still operate profitably despite the penalty in efficiency imposed by CCS. They must be in close proximity to a site with suitable geological storage, and they must have a supportive policy environment.

A set of threshold criteria might be chosen to assess a power station's 'retrofitability' with CCS: nameplate generating capacity in excess of 300 MW, carbon intensity less than 900 $kgCO_2$ *per* MWh as a proxy of efficiency, less than 20 years of age, 40 km to suitable geological storage, and a 'high' policy support rating from the Global CCS Institute. With only few

Table 7 Projected deployment of Carbon Capture and Storage (CCS) under various low-carbon scenarios. (Data drawn from International Energy Agency[94,95] and the Intergovernmental Panel on Climate Change.[7])

Scenario	Projected deployment
IEA World Energy Outlook: 450S	5 $GtCO_2$ per year, 60% in power
IEA Energy Technology Perspectives: 2DS	4 $GtCO_2$ per year, 57% in power
IPCC AR5: Cost-efficient 430–530 ppm	5.5 to 12.1 $GtCO_2$ per year (imputed)

exceptions, less than 25% of the coal-fired generating capacity of companies in the top 100 coal-fired power utilities would be considered 'retrofitable' by these criteria. Retrofitability does not seem to have been a guiding factor in the choice to build a new power station, and for most power stations retrofitability is not an option for avoiding stranded assets. See Caldecott *et al.*[22,99] for further analysis.

There are challenging political economy implications of the direct policy support that CCS-equipped fossil power generation requires in order to reduce its costs to a level competitive with other decarbonised generating options. First, the scale of subsidy required to incentivise CCS capital programmes requires that policy makers 'pick winners' among clean technology options and individual projects – an uncomfortable prospect for governments that prefer auction and market mechanisms for efficient subsidy allocation. The short-term inefficient allocation of public funds to CCS projects (relative to other lower-cost mitigation options) is justified by the potential long-term cost reductions; however, these reductions are uncertain at best. Second, policymakers and the public may be uncomfortable subsidising the same companies to capture and store carbon dioxide who have been the historic benefactors of the socialised externality of carbon emissions. Third, CCS enables the inclusion of human capital locked into high-emitting industries; however, it may slow the de-legitimisation and transition of the same.

4.8 Availability of Finance

Public pressure has slowed the development of carbon-intensive electricity generation in favour of low-carbon technologies, and it has particularly targeted coal through the divestment movement. Many bilateral and multilateral financing organisations responsible for energy infrastructure projects in developing countries, most of whom are seen as the future developers of coal power projects, have announced funding limitations for coal power projects. Announcements from the European Investment Bank, the World Bank Group and the Organization for Economic Co-operation and Development have, to different degrees, put forth standards that will restrict the financing of coal power projects to only extreme circumstances where they deem that there are no feasible alternatives to meet people's basic energy needs. When development banks began restricting multi-lateral finance for coal projects, national export credit agencies filled the financing gap, and more than half of recent investment came from export-import banks, which provide government-backed loans, credits and guarantees from home countries to back higher-risk projects. But the scale of investment from multilateral finance institutions for power generation projects is not exactly significant: in 2013, multilateral institutions only invested in less than 10% of commissioned projects in India.[100]

China has grown to become a major, if not the largest, source of financing for coal in overseas markets.[101] Over the past ten years, China's financing institutions provided between US$21bn and US$38bn, with another US$35–72bn in their pipeline; China "has not only been stepping in to replace the market share of entities and countries who pledged to restrict support to coal power, but also the share of funding countries who have not pledged to do so".[101] Commercial banks provided at least $500bn over the past decade (and this is probably an underestimate due to measurement difficulties), significantly overshadowing public financing numbers.[102] China's banks are rapidly increasing their market share: in the past three years, the make-up of the top financing institutions has changed significantly, three of the top six private banks funding coal projects are headquartered in China. Western banks, including HSBC, Deutsche Bank, Citi and Morgan Stanley, have divested.[102]

Despite the abundance of political, financial and reputational downside risk, the reality remains that government-sponsored coal projects in, for example, India continue to attract sufficient funding to develop.[103] Indian Prime Minister Narendra Modi and Energy Minister Piyush Goyal have promised to develop energy at an unprecedented scale, including at least 200 GW of new thermal generation capacity by 2047 and a doubling of domestic coal use from 565 million tons to over one billion tons by 2019.[104] If India does follow its intended path, relying heavily on fossil fuels to increase standards of living and growing the economy, then perhaps "the mitigation actions of the US and China won't matter".[105]

5   Conclusion

Power generation is the primary end-use of coal; the future of coal in the 21st century is largely the future of coal-fired power. New high temperature and pressure steam generator designs are pushing the limit of power station efficiency in order to reduce greenhouse gas and conventional air pollutants. Some novel technological configurations like IGCCs, coal liquefaction and underground coal gasification offer some interesting new prospects for coal by alleviating certain environmental drawbacks and unlocking arbitrage with liquids and gas energy carriers. However, many of these technologies suffer from the same environmental challenges as conventional coal-fired power.

Coal-fired power is the single largest sector of global emissions and has ready technical replacements in the form of nuclear, gas and renewable power. The elimination of coal-fired power is a low-hanging fruit for policy makers seeking to limit climate change to safe levels of warming. Multiple authors and studies find the immediate reduction and elimination of coal-fired power option is the most efficient pathway to a 2 °C-constrained future. Coal-fired power has a large water consumption footprint, which will only be exacerbated by future climate stresses. Conventional air pollution emissions from coal-fired power have already caused crises in China and India. Coal-fired power stations are unprepared to meet the flexibility requirements of a future grid abundant with renewables. Options to decarbonise coal-fired power generation like CCS are expensive and not forthcoming. Ownership of coal assets is becoming a reputational risk as the concerned public begins to engage with financial institutions on their financing of climate-destroying coal projects.

Coal is an inexpensive and sovereign resource for many countries. The promise of coal to deliver global electrification is double-edged: many of the same countries considering coal for their development plans are the most vulnerable to changes in the climate. A prosperous future ultimately requires clear air, clean water, and a stable climate. Coal-fired power finds itself at a salient juncture with rapid decline on one side and indeterminate growth on the other. The outcome will be determined by the response of policy makers, investors, companies and the public to a wide range of environmental risks.

References

1. International Energy Agency, *World Energy Outlook 2016*, International Energy Agency, Paris, France, 2016.
2. International Energy Agency, *Medium-Term Coal Market Report 2016*, International Energy Agency, 2016.
3. International Energy Agency, *World Energy Model Documentation*, International Energy Agency, 2016.
4. MIT CCS Project Database as of September 30, 2016. [Internet] 2016. Available from: https://sequestration.mit.edu/tools/projects/index_capture.html. Accessed 21 Apr 2017.
5. International Energy Agency, Special Report: World Energy Investment Outlook, International Energy Agency, Paris, France, 2014. [Internet] 2014.
6. A. Pfeiffer, R. Miller, C. Hepburn and E. Beinhocker, *Appl. Energy*, 2016, **179**, 1395.
7. IPCC, *Climate Change 2014: Mitigation of Climate Change. Contribution of Working Group III to the Fifth Assessment Report of the Intergovernmental Panel on Climate Change*, ed. O. Edenhofer, R. Pichs-Madruga, Y. Sokona, E. Farahani, S. Kadner, K. Seyboth, A. Adler, I. Baum, S. Brunner, P. Eickemeier, B. Kriemann, J. Savolainen, S. Schlömer, C. von Stechow, T. Zwickel and J. C. Minx, Cambridge University Press, Cambridge, United Kingdom and New York, NY, USA, 2014.
8. J. Rogelj, G. Luderer, R. C. Pietzcker, E. Kriegler, M. Schaeffer, V. Krey and K. Riahi, *Nat. Clim. Change*, 2015, **5**, 519.
9. B. Anthony, CCS and CCUS, in *Coal in the 21st Century: Energy Needs, Chemicals, and Environmental Controls*, ed. R. E. Hester and R. M. Harrison, Royal Society of Chemistry, Cambridge, UK, 2017, ch. 7.
10. J. Salvatore, *World Energy Perspective: Cost of Energy Technologies*, World Energy Council, London, UK, 2013.
11. C. Shearer, N. Ghio, L. Myllyvirta, A. Yu and T. Nace, Boom and Bust 2017: Tracking the Global Coal Plant Pipeline, Coalswarm/Sierra Club/Greenpeace, 2017.

12. J. W. Leonard, *Coal Preparation*, Society for Mining, Metallurgy, & Exploration, Englewood, CO, 5th edn, 1991.
13. S. Nunes, *Coal Upgrading*, IEA Clean Coal Centre, London, UK, 2009.
14. M. Kilma, B. Arnold, J. Barbara and P. J. Bethell, *Challenges in Fine Coal Processing, Dewatering, and Disposal*, Society for Mining, Metallurgy, & Exploration, Engelwood, CO, 2012.
15. M. Susta and K. B. Seong, Supercritical and Utra-Superciritical Power Plants – SEA's vision or reality?, Powergen Asia, 2004.
16. P. R. Shukla, A. Rana, A. Garg, M. Kapshe and R. Nair, *Climate Policy Assessment for India*, Universities Press, Hydrabad, India, 2004.
17. F. Wong, China to ban imports of high ash, high sulphur coal from 2015. Reuters. [Internet] 2014. Available from: http://uk.reuters.com/article/us-china-coal-imports-idUSKBN0HB02M20140916. Accessed 5 Apr 2017.
18. C. Higman and M. van der Burgt, *Gasification*, Gulf Professional Publishing, 2nd edn, 2008.
19. R. Fernando, *Coal Gasification*, IEA Clean Coal Centre, London, UK, 2008.
20. Y. Guo, Z. Huang and Z. Zhou, *Energy Power Eng*, 2015, **7**(11), 535.
21. C. Snape, Liquid fuels and chemicals/plastics from coal, in *Coal in the 21st Century: Energy Needs, Chemicals, and Environmental Controls*, ed. R. E. Hester and R. M. Harrisson, Royal Society of Chemistry, Cambridge, UK, 2017, ch. 6.
22. B. Caldecott, L. Kruitwagen, G. Dericks, D. J. Tulloch, I. Kok and J. Mitchell, Stranded Assets and Thermal Coal: An Analysis of Environment-related Risks, Smith School of Enterprise and the Environment, Working Paper, University of Oxford, Oxford, UK, 2016.
23. F. Mao, *Nat. Gas Ind. B*, 2016, **4**(3), 312.
24. R. Anderson, Coal gasification: The clean energy of the future? *BBC News*, London, UK, 2014.
25. G. Chen, Clean Coal Technology for Advanced Power Generation, *Research Center for Energy Technology and Strategy*, Presentation: 8 August 2014.
26. C. Higman, State of the Gasification Industry: Worldwide Gasification Database 2014 Update, Gasification Technologies Conference, 29 October 2014, [Internet] 2014. Available from: http://www.gasification-syngas.org/uploads/downloads/GTC_Database_2014.pdf. Accessed 23 Apr 2017.
27. J. Tennant, Gasification Systems Overview, US Department of Energy, Presentation, [Internet] 2016. Available from: https://www.netl.doe.gov/File%20Library/Research/Coal/energy%20systems/gasification/doe-gasification-program-overview.pdf. Accessed 23 Apr 2017.
28. National Energy Technologies Laboratory (NETL). Coal Gasification and Transportation Fuels Quarterly News. 1:1-4, 2:1-3. [Internet] 2016. Available from: https://www.netl.doe.gov/research/coal/energy-systems/gasification/coal-gasification-magazine. Accessed 23 Apr 2017.

29. P. Breeze, *Coal-Fired Generation*, Academic Press, London, 1st edn, 2015.
30. B. G. Miller, *Clean Coal Engineering Technology*, Butterworth-Heinemann, 2010.
31. J. Katzer, *The Future of Coal: Options for a Carbon-Constrained World*, MIT, 2007.
32. T. Riordan, New Generation Strategy Ultra-Supercritical Technology, AEP, Presentation, [Internet] 2006. Available from: https://view. officeapps.live.com/op/view.aspx?src=. Available from: http://www. asiapacificpartnership.org/pdf/PGTTF/events-october-06/24%20-% 20New%20Gen%20Strategy%20Ultra-Supercritical%20Technlgy.ppt. Accessed 11 Apr 2017.
33. European Commission, Advanced Fossil Fuel Power Generation Thematic Research Summary, European Union, 2014.
34. J. Phillips and J. Wheeldon, Economic Analysis of Advance Ultra-Supercritical Pulverized Coal Power Plants: A Cost-Effective CO2 Emission Reduction Option? Electric Power Research Institute, Presentation: Advances in Materials Technology for Fossil Power Plants 6th International Conference, Santa Fe, New Mexico, [Internet] 2011. Available from: http://www.asminternational.org/documents/10192/ 3298473/cp2010epri0053.pdf/49316295-9452-45fd-a428-6eaad5d7ab91. Accessed 11 Apr 2017.
35. Benson Boiler Technology, *Siemens*, 2009.
36. B. Caldecott, G. Dericks, D. J. Tulloch, X. Liao, L. Kruitwagen, G. Bouveret and J. Mitchell, Stranded Assets and Thermal Coal in China, Smith School of Enterprise and the Environment, Working Paper, University of Oxford, Oxford, UK, 2017.
37. Gasification Plant Databases: June 2016 Update, National Energy Technology Laboratory, [Internet] 2016. Available from: https://www. netl.doe.gov/research/coal/energy-systems/gasification/gasification-plant-databases. Accessed 23 Apr 2017.
38. G. Chen, Clean Coal Technology for Advanced Power Generation, Research Center for Energy Technology and Strategy, Presentation: 8 August 2014.
39. P. Wang, N. Means, D. Shekawat, D. Berry and M. Massoudi, *Energies*, 2015, **8**, 10605.
40. A. Cuadrat, A. Abad, F. Garcia-Labiano, P. Gayan, L. F. de Diego and J. Adanez, *Chem. Eng. J.*, 2012, **195**, 91.
41. L. G. Velazquet-Vargas, D. J. Devault, T. J. Flynn, T. Siengchum, L. Zeng, A. Tong, S. Bayham and L. S. Fan, Techno-economic analysis of a 550 MWe atmospheric iron-based coal-direct chemical looping process, in *Proceeding of 3rd International Conference on Chemical Looping*, Goteborg, Sweden, 9–11 September 2014.
42. K. Kimmel and R. Cleetus, Regulation, in *Coal in the 21st Century: Energy Needs, Chemicals, and Environmental Controls*, ed. R. E. Hester and R. M. Harrison, Royal Society of Chemistry, Cambridge, UK, 2017, ch. 5.

43. C. McGlade and P. Ekins, *Nature*, 2015, **517**, 187.

44. The Carbon Bubble, *Carbon Tracker Initiative*, 2011.

45. R. Stefanski, Dirty Little Secrets: Inferring Fossil-Fuel Subsidies from Patters in Emissions Intensities, OxCarre Research Paper 134, 2015.

46. B. Caldecott, N. Howarth and P. McSharry, Stranded Assets in Agriculture: Protecting Value from Environment-related Risks, Smith School of Enterprise and the Environment, Working Paper, University of Oxford, Oxford UK, 2013.

47. B. Caldecott, Avoiding Stranded Assets, in *Confronting Hidden Threats toSustainability*, ed. G. Gardner, Worldwatch Institute, Washington, USA, 2015, ch. 4.

48. B. Caldecott and J. McDaniels, Financial Dynamics of the Environment: Risks, Impacts, and Barriers to Resilience, Smith School of Enterprise and the Environment, Working Paper, University of Oxford, Oxford, UK, 2014.

49. M. Carney, Breaking the tragedy of the horizon – climate change and financial stability, Speech given at Lloyds of London, Bank of England, 2015.

50. TCFD Recommendations Report, Task Force on Climate-Related Financial Disclosures, 2016.

51. Water Use for Electric Power Generation, Electric Power Research Institute (EPRI), Palo Alto, CA, 2008, 1014026.

52. J. Meldrum, S. Nettles-Anderson, G. Heath and J. Macknick, *Environ. Res. Lett.*, 2013, **8**, 18.

53. M. Kurian, *J. Environ. Sci. Policy*, 2017, **68**, 97.

54. D. Feeny, F. Berkes, B. J. McCay and J. M. Acheson, *Hum. Ecol.*, 1990, **18**(1), 1.

55. International Energy Agency, *World Energy Outlook 2012*, International Energy Agency, Paris, France, 2012.

56. L. Sussams, The Great Coal Cap: China's energy policies and the financial implications for thermal coal, Carbon Tracker Initiative, 2014.

57. B. Gottlieb and A. Lockwood, Health Impacts, in *Coal in the 21st Century: Energy Needs, Chemicals, and Environmental Controls*, ed. R. E. Hester and R. M. Harrison, Royal Society of Chemistry, Cambridge, UK, 2017, ch. 4.

58. D. Vallero, *Fundamentals of Air Pollution*, Academic Press, London, 5th edn, 2014.

59. X. Zhang, *Management of coal combustion wastes*, IEA Clean Coal Centre, 2014.

60. Health Effects Institute, State of Global Air 2017: A Special Report on Global Exposure to Air Pollution and its Disease Burden, Health Effects Institute, 2017.

61. K. Deep Singh, What Delhi Government Says it is Doing to Combat Air Pollution, *The Wall Street Journal*, [Internet] 2016. Available from: https://blogs.wsj.com/indiarealtime/2016/11/07/what-delhi-government-says-it-is-doing-to-combat-air-pollution/. Accessed 11 Apr 2017.

62. O. Aydin, F. Graves and F. Celebi, Coal Plant Retirements: Feedback Effects on Wholesale Electricity Prices, *The Brattle Group*, 2013.
63. K. Haustein, F. Otto, P. Uhe, N. Schaller, M. Allen, L. Hermanson, N. Christidis, P. McLean and H. Cullen, *Environ. Res. Lett.*, 2016, **11**, 1.
64. F. Otto, R. James and M. Allen, The science of attributing extreme weather events and its potential contribution to assessing loss and damage associated with climate change impacts, Oxford Environmental Change Institute, Briefing Note for UNFCC, 2016.
65. F. Otto, J. Geert, J. Eden, P. Stott, D. Karoly and M. Allen, *Nat. Clim. Change*, 2016, **6**, 813.
66. C. Funk, G. Husak, J. Michaelsen, S. Shukla, A. Hoel, B. Lyon, M. P. Hoerling, B. Liebmann, T. Zhang, J. Verdin, G. Galu, G. Eilerts and J. Rowland, *Bull. Am. Meteorol. Soc.*, 2013, **94**(9), S6.
67. F. Lott, N. Christidis and P. Stott, *Geophys. Res. Lett.*, 2013, **40**, 1177.
68. S. Herring, A. Hoell, M. Hoerling, J. Kossil, C. Schreck III and P. Stott, *Bull. Am. Meteorol. Soc.*, 2016, **97**, 12.
69. S. Herring, A. Hoell, M. Hoerling, J. Kossil, C. Schreck III and P. Stott, *Bull. Am. Meteorol. Soc.*, 2015, **96**, 12.
70. M. Light and S. Cedres, Impact of Long-Term Drought on Power Systems in the U.S. Southwest., US DOE Office of Electricity Deliver and Energy Reliability Infrastructure Security and Energy Restoration Division, 2012.
71. J. Kern, Financial vulnerability of the electricity sector to drought, and the impacts of changes in generation mix, American Geophysical Union, Abstract #H34F-03, 2015.
72. M. A. Maupin, J. F. Kenny, S. S. Hutson, J. K. Lovelace, N. L. Barber and K. S. Linsey, Estimated use of water in the United States in 2010, U.S. Geological Survey Circular 1405, 2014.
73. J. Eyer and C. Wichman, Does Water Scarcity Shift the Electricty Generation Mix toward Fossil Fuels? Emperical Evidence from the United States, Resources for the Future, 2016.
74. S. Dharmandhikary and S. Dixit, Thermal Power Plants on the Anvil: Implications and Need for Rationalisation, Prayas Energy Group, Discussion Paper, 2012.
75. S. Dharmandhikary, Thermal Power Plans: Giant Water Guzzler in Drought-Ridden India, Ecologise.in. [Internet] 2016. Available from: http://www.ecologise.in/2016/06/10/thermal-power-plants-giant-water-guzzlers-in-drought-ridden-india/. Accessed 11 Apr 2017.
76. I. Cheng and H. Lammi, The Great Water Grab: How the Coal Industry is Deepening the Global Water Crisis, *Greenpeace*, 2016.
77. K. Schneider, As Drought Grips South Africa, A Conflict Over Water and Coal, Yale School of Forestry & Environmental Studies, [Internet] 2016. Available from: http://e360.yale.edu/features/south_africa_drought_coal_renewables. Accessed 19 Apr 2017.
78. M. Van Vliet, D. Wiberg, S. Leduc and K. Riahi, *Nat. Clim. Change*, 2015, **6**, 375.

79. J. Kern, Financial vulnerability of the electricity sector to drought, and the impacts of changes in generation mix, *American Geophysical Union, Fall Meeting 2015 Abstract #H34F-03*, 2015.

80. S. Aivalioti, Electricity Sector Adaptation to Heat Waves. *Sabin Center for Climate Change Law*, Columbia Law School, 2016.

81. M. A. Cook, C. W. King, F. T. Davidson and M. E. Webber, *Energy Rep.*, 2015, **1**, 193.

82. National Energy Technology Laboratory, Cost and Performance Baseline for Fossil Energy Plants Volume 3c: Natural Gas Combined Cycle at Elevation, U.S. Department of Energy, DOE/NETL-2010/1396, 2011.

83. T. Fout, A. Zoelle, D. Kearins, M. Turner, M. Woods, N. Keuhn, V. Shah, V. Chou and L. Pinkerton, Cost and Performance Baseline for Fossil Energy Plants Volume 1a: Bituminous Coal (PC) and Natural Gas to Electricity Revision 3, US Department of Energy Office of Fossil Energy, DOE/NETL-2015/1723, 2015.

84. National Energy Technology Laboratory, Cost and Performance Baseline for Fossil Energy Plants Volume 3b: Low Rank Coal to Electricity: Combustion Causes, *US Department of Energy*: DOE/NETL-2001/1463, 2011.

85. K. Appunn, Setting the power price: the merit order effect. *European Commission Directorate-General for Energe (ECDGE)*, ECDGE Staff Working Document, SWD/2015/0142, 2015.

86. G. Strbac, Workshop on Addressing Flexibility in Energy System Models, European Commission Strategic Energy Technologies Information System (SETIS), [Internet] 2015. Available from: https://setis.ec.europa.eu/system/files/Slides%20-%2017%20Strbac%20%28ICL%29.pdf. Accessed 11 Apr 2017.

87. D. Lunn, Technical Assessment of the Operation of Coal & Gas Fired Plants, *Parsons Brinckerhoff*, 206861A: Prepared for the UK Department of Energy and Climate Change, 2014.

88. N. Kumar, P. Besuner, S. Lefton, D. Agan and D. Hilleman, *Power Plant Cycling Costs*, National Renewable Energy Laboratory, 2012.

89. Cost and Performance Characteristics of New Generating Technologies, Annual Energy Outlook 2017, *Energy Information Administration (EIA)*, 2017. [Internet] 2017.

90. M. Sontakke, Nuclear Power Plants are cheaper to operate, Market Realist, [Internet] 2015. Available from: http://marketrealist.com/2015/01/nuclear-power-plants-cheaper-operate/. Accessed 22 Apr 2017.

91. National Energy Technology Laboratory (NETL), Cost and Performance Baseline for Fossil Energy Plants Volume 1: Bituminous Coal and Natural Gas to Electricity, *US Department of Energy*, DOE/NETL-2010/1397, 2013.

92. T. Stacey and G. Taylor, The Levelized Cost of Electricity from Existing Generation Resources, Institute for Energy Research, 2015.

93. National Energy Technology Laboratory (NETL). Cost and Performance Baseline for Fossil Energy Plants Volume 1: Bituminous Coal and

Natural Gas to Electricity. US Department of Energy, DOE/NETL-2010/1397, 2013.

94. International Energy Agency, *World Energy Outlook 2015*, International Energy Agency, Paris, France, 2015.

95. International Energy Agency, *Technology Roadmap: Carbon Capture and Storage*, International Energy Agency, Paris, France, 2013.

96. Large Scale CCS Projects, Global CCS Institute, [Internet] 2017. Available at: https://www.globalccsinstitute.com/projects/large-scale-ccs-projects. Accessed 1 Apr 2017).

97. K. Gerdes, Incentivizing Carbon Capture Retrofits of the Existing PC and NGCC Fleet, National Energy and Technology Laboratory, Presentation, [Internet] 2014. Available from: https://www.netl.doe.gov/File%20Library/Events/2014/2014%20NETL%20CO2%20Capture/K-Gerdes-NETL-Incentivizing-Carbon-Capture-Retrofits.pdf. Accessed 12 Apr 2017.

98. M. Finkerath, J. Smith and D. Volk, *CCS Retrofit: Analysis of the Globally Installed Coal-Fired Power Plant Fleet*, International Energy Agency, 2012.

99. B. Caldecott, L. Kruitwagen and I. Kok, *Oxford Energy Forum*, 2016, **105**, 50.

100. Ministry of Power, Annual Report 2012–2013, Government of India, Ministry of Power, 2013.

101. M. Herve-Mignuccia and X. Wang, Slowing the Growth of Coal Power Outside China: The Role of Chinese Finance, *Climate Policy Initiative*, 2015.

102. Y. Louvel, R. Brightwell and G. Aitken, Banking on Coal 2014: Bank Financing of Coal Mining and Coal Power, *BankTrack*, 2014.

103. T. Buckley, India's Electricity-Sector Transformation, *Institute for Energy Economics and Financial Analysis*, 2015.

104. India Energy Security Scenarios (IESS) 2047, User Guide for India's 2047 Energy Calculator Coal and Gas Power Stations, *Indiaenergy.gov* 2015, [Internet]. Available from: http://indiaenergy.gov.in/docs/Thermal-power-generation-documentation.pdf. Accessed 9 Apr 2017).

105. R. Martin, India's Energy Crisis, MIT Technology Review [Internet] 2015. Available from: http://www.technologyreview.com/featuredstory/542091/indias-energy-crisis/. Accessed 19 Apr 2017.

The Life Cycle of Coal and Associated Health Impacts

BARBARA GOTTLIEB* AND ALAN LOCKWOOD

ABSTRACT

Coal-related pollution makes important contributions to four of the five leading causes of death in the US: heart disease, cancer, chronic lower respiratory diseases and stroke. Diabetes mellitus and Alzheimer's disease may join that list. Every stage of the so-called mine-to-waste "life cycle" of coal use is associated with its own and often unique threats to health. Critical stages include mining, transporting coal from the mine to the site where it is burned, hazardous air pollutants released during combustion, and the disposal of coal combustion waste, often referred to as coal ash. In addition, the carbon dioxide released by burning coal and other fossil fuels is the single most important cause of climate change. Climate change is seen by some as either the greatest threat to humankind in this century or, as it has been called more optimistically, "the greatest public health opportunity of the 21st century".

1 Introduction

In spite of growing evidence that using coal as an energy source to produce electricity poses serious health threats, coal continues to be central to the world's expanding energy needs. The most recent data from the International Energy Institute indicate that coal supplies around 41% of this energy. And, the percentage is increasing. This is in response to the growing

---

*Corresponding author.

Issues in Environmental Science and Technology No. 45
Coal in the 21st Century: Energy Needs, Chemicals and Environmental Controls
Edited by R.E. Hester and R.M. Harrison
© The Royal Society of Chemistry 2018
Published by the Royal Society of Chemistry, www.rsc.org

demand for electricity in countries as they strive to develop their economies and to become more prosperous. It is a basic fact: modern civilizations cannot exist, let alone grow, without large amounts of electricity.

Every aspect of coal utilization poses health threats, ranging from accidents and black lung disease (more properly known as coal workers' pneumoconiosis or CWP) in miners to the inhabitants of the central Pacific island nation of Kiribati who have become climate refugees due to rising sea level caused by climate change.

2 Mining

2.1 Accidents

At 3:27 in the afternoon on April 5, 2010, an explosion ripped through Massey Energy's Upper Big Branch Mine in Raleigh County, West Virginia, killing 29 of the 31 miners who were working in the mine. A report was published which examined this, the worst US mine disaster since 1970.[1] As described in the report, which included details of the subsequent investigation by the Mine Safety and Health Administration, 369 citations and orders were issued to the mine operator, Performance Coal Company, a subsidiary of Massey Energy Company. The Agency also imposed a record fine of $10 825 368. The report concluded that the root cause of the "accident" was Massey's corporate culture. We place *accident* in quotation marks to draw attention to the fact that this was not an accident in the traditional sense of the word. It was a disaster due to negligent operation of the mine. The report includes a statement by then-Secretary of Labor Hilda Solis, who said the mine owners and operators "promoted and enforced a workplace culture that valued production over safety, and broke the law as they endangered the lives of their miners." Subsequently, Donald Blankenship, who led Massey Energy at the time, was sentenced to a year in prison for his role in the disaster.

Mining coal is dangerous. According to a 2014 report by the US Bureau of Labor Statistics (BLS), US workers with jobs in "mining, quarrying, and oil and gas extraction" had the second-highest fatality rate among private industries.[2] The fatality rate was 14.2 deaths *per* 100 000 full-time equivalent workers employed 40 hours *per* week, 50 weeks *per* year. By contrast, the rate for all government employees was 1.9 deaths *per* 100 000 employees. When restricted to just coal miners, the most recent data list 28 fatal injuries or 24.8 *per* 100 000 full-time employees, a 57% decrease from the prior year. Although this apparent decrease was encouraging, there is a substantial amount of variability among these data due to the fact that the total number of coal miners and accidents is relatively small compared to the total size of the US work force, and year-to-year variability is the rule rather than the exception. Some of this variability is due to the fact that single accidents may claim a substantial number of lives, as was the case at the Big Branch Mine.

Among the 2007 deaths, 71% were sustained by miners in underground bituminous coal mines. "Contact with objects and equipment" along with transportation injuries were the most common types of fatal accidents. By comparison, 375 workers were shot to death while on the job in 2012. Even though the risk of being shot to death at work is much lower than accidental death among coal miners, the huge size of the total work force inflates the apparent risk of being shot while at work.

According to the BLS report, non-fatal injuries among coal miners were much more common, occurring at a rate of 4.4 *per* 100 full-time employees. Again, underground bituminous coal mines posed the greatest risk. Thus, the risk for injury among underground miners is orders of magnitude higher than the risk for death. Both are unacceptably high.

2.2   Coal-worker's Pneumoconiosis (CWP) or Black Lung Disease

Coal-worker's pneumoconiosis (CWP), commonly known as Black Lung Disease, is a chronic disease of the lungs. This occupational disease is caused by the sustained inhalation of coal dust.[3] The finest particles of dust enter the lungs and travel into the alveoli, the site where the exchange of carbon dioxide for oxygen occurs. The alveoli contain cells (macrophages) that ingest the dust particles. These macrophages tend to collect and form coal macules, one of the lesions that is characteristic of CWP. The macrophages trigger a series of responses that result in irritation and inflammation and enlargement of the portions of the respiratory tree called bronchioles. With enough time and exposure, these inflammatory changes cause the formation of fibers in the lungs (fibrosis) that impair lung function. This marks the beginnings of chronic bronchitis. In the more extreme cases, air spaces in the lung enlarge. Centrilobular emphysema is the result of this enlargement. When coal macules coalesce and one or more reaches 2 cm in diameter, the condition is properly labeled as progressive massive fibrosis (PMF). This is the most severe form of CWP.

Although what we now call CWP is thought to have been known to the Chinese before the Common Era, one of the early descriptions of the disease was offered by Archibald Makellar in his 1846 monograph titled "An Investigation into the Nature of Black Phthisis: or Ulceration Induced by Carbonaceous Accumulation in the Lungs of Coal Miners and other Operatives." He wrote:[4]

> "A robust young man, engaged as a miner, after being for a short time so occupied, becomes affected with cough, inky expectoration, rapidly decreasing pulse, and general exhaustion. In the course of a few years, he sinks under the disease; and, on examination of the chest after death, the lungs are found excavated, and several of the cavities filled with a solid or fluid carbonaceous matter."

Makellar was describing what is now called PMF.

Risk factors among miners that predispose them to the development of CWP are: (i) working underground and (ii) the number of years spent in the mining industry.[5] In a federally administered CWP chest X-ray surveillance program, the overall incidence of CWP in X-ray films from a cohort of almost 36 000 was 2.8% and the incidence of PMF was 0.2%. Among miners less than 30 years of age, the prevalence of CWP was 0.2%, rising with age to 5.1% among those 60 years old or older. Underground miners had a higher prevalence than surface miners, as did those working in mines with 50 workers or less.

In 1969 the US Congress passed the Coal Mine Safety and Health Act. This act had two major provisions. It established maximum coal dust exposure levels and a voluntary chest X-ray screening program. As a result, the prevalence of CWP fell in the 1970s, only to show a rise in the 1990s.[6] This was particularly true among workers in mines employing 50 or fewer miners where the prevalence was five times greater than among miners working in large mines.

As a response to these and other data, new regulations designed to protect miners were promulgated in 2014.[7] These regulations expanded the medical surveillance of miners by including symptom assessment and tests of pulmonary function. Importantly, it lowered the limit for respirable dust concentration and mandated the use of continuous personal dust monitoring devices to aid in the minimization of coal dust exposure. By providing real-time exposure data, monitoring devices allow for the working environment to be adjusted quickly to protect the health of miners. They are another link in the exposure-avoidance practices needed to reduce the risk for developing CWP and are an example of how primary prevention can benefit workers.

At about this same time, and ironically on the 45th anniversary of the Federal Coal Mine Health and Safety Act, the National Institutes of Safety and Health (NIOSH) noted a sharp increase in the prevalence of PMF.[8] According to the monitoring programs in effect at the time, the disease had a prevalence of 0.08% among all participants and 0.33% among underground miners with 25 years or more experience. A five-year moving average of prevalence data showed nearly a ten-fold increase (note: a five-year moving average is a statistical technique designed to smooth out spikes or troughs in data by averaging data from a given year with data from the two preceding and the two following years).

The wisdom of the 2014 rule was reinforced by a 2016 observation made by a single radiologist in Kentucky.[7] Between January of 2015 and August of 2016 this radiologist, who was board-certified and had received special training and certification in the interpretation of X-rays in order to diagnose CWP, identified 60 cases of PMF in current and former miners. The interpretations he made were confirmed by others with similar training and experience. Substantial numbers of the affected miners worked as roof bolters, the miners whose job it is to insert bolts into the roof of mine shafts to prevent their collapse, and miners who operated continuous mining machines that loosen coal along with some adjacent rock at the mine face.

CWP is a severe disease without any effective treatment, other than supportive care. It is entirely preventable. Thus, stringent measures to protect the health of miners are warranted to stem this epidemic.

2.3 Coal Mining Effects on Water Quality

Finally, coal mines have adverse impacts on water quality in nearby streams.[9] In 2012, a group of university-based investigators performed a systematic study of waterways in a 19 581 km^2 area of West Virginia between 1976 and 2005. They found that in regions where surface mines occupied 5.4% or more of the surface area, pollutant levels in streams exceeded levels known to cause biological impairment. For particularly sensitive organisms, the threshold for impairment was reached when mines occupied 2.2% of the surface area. They also reported that 5% of the land area of southern West Virginia was covered by surface mines, frequently of the mountaintop-removal and valley-fill variety. Valley fills buried 6% of regional streams, and 22% of the stream network drained regions where the 5.4% mine area threshold had been exceeded. These findings have notable implications for water quality and the diversity of species in areas where coal is mined.

2.4 Effects of Mining on the Health of Residents in Adjacent Communities

At about the same time as it was becoming evident that CWP was on the rise, a series of reports began to draw attention to the health risks associated with living in regions where there was a substantial amount of coal mining.

An important early study reported on the results of a survey of over 16 000 residents of West Virginia that was merged with county-wide coal mining data.[10] Mining data were used to divide counties where 4 million tons of coal were mined each year from those where less coal was mined. The results showed that respondents living in the high-mining counties reported significantly poorer overall health, compared to residents where less than four million tons were mined annually. Similarly significant relationships were found between mine proximity and any lung disease (presumably including CWP), hypertension, and kidney disease (possibly due to an association with high blood pressure). Additional statistical analyses showed that the probability that one had chronic obstructive pulmonary disease (COPD) was around 50% higher among those living in counties where 4 million or more tons of coal were mined annually.

Mortality rates are also elevated in coal mining counties throughout Appalachia.[11] In this national countywide study of data compiled for the years between 1999 and 2004, the overall age-adjusted mortality rates were significantly elevated where the largest amount of coal was mined. This did not appear to be an all-or-nothing phenomenon, as mortality rates increased steadily as coal extraction rose from 1 to 7 million tons *per* year. This effect was present even after controlling for various potentially confounding variables such as smoking, poverty, education, rural *versus* urban settings, and others. The authors concluded that there were just over 1600 excess annual deaths in Appalachian coal mining areas due to mine proximity.

The authors of the above studies on health effects in Appalachia conclude that the weight of the evidence in the literature suggests that this burden of disease is due to the toxins released into the environment. Coals do not have a uniform composition and contain a large number of hydrocarbons and elements that are linked to disease. Some are released during combustion, as discussed below, and others have the potential for release during the mining process itself as well as washing and other steps used to prepare coal for shipping. It is hypothesized that these toxins act by pathways associated with inflammatory responses.[1]

3 Coal Preparation and Transport

3.1 Coal Cleaning; Mining Wastes

Virtually all of the coal that is mined contains rock and other impurities that contaminate the coal. As mining has proceeded, particularly in the Appalachian coal fields, high-quality coal has been depleted and residual deposits contain more impurities or are more difficult to mine. Coal cleaning operations are designed to remove some of this unwanted non-coal material referred to as "partings". Depending on the specific properties of a mine, between 20% and 50% of the extracted material is rejected and diverted into a waste stream.[12] Purification mechanisms include: crushing, screening (sorting by size) and washing. Many of the washing strategies rely on differences in the physical properties of coal and impurities, such as the density or weight *per* unit volume of coal *versus* rock. Substantial amounts of water are required for many of these mechanisms. Large centrifuges or liquids with differing densities separate the less-dense coal from the more-dense rock. These cleaning processes make coal less expensive to ship and yield a final product that has more consistent properties. This consistency is a critical factor in the combustion processes. These processes gave rise to the initial concept of "clean coal," a term that now has a different use, referring to carbon dioxide capture and storage. The downside of cleaning mechanisms lies with the substantial amount of waste that must be disposed of. Strategies that consume water produce a slurry that is a combination of waste water and partings. Slurry is frequently stored in ponds behind dams.

3.2 Coal Waste Accidents

The perils associated with coal-washing sludge were carved into the memories of West Virginians on February 26, 1972.[13] That morning a worker at the Buffalo Creek Mine noticed that the highest of three successive dams built to restrain the coal-washing slurry looked "real soggy." Within minutes, the dam collapsed. The impounded slurry raced down Buffalo Hollow Creek, overwhelming the remaining two lower dams. An estimated 132 million gallons of water and waste inundated the town, killing 125 and injuring another 1100. The number of those rendered homeless was around 4000. It remains as one of the worst flooding disasters in the history of the US. The lawsuits that followed resulted in the payment of around $13 000 to each of the plaintiffs. The power of the coal companies is illustrated by a $1 million settlement after the State sued the coal company for $100 million.

Another well-publicized sludge spill occurred on October 11, 2000, when the bottom of a 72-acre waste impoundment lake in Martin County, Kentucky, owned by Massey Energy, collapsed into an underground mine. Over 300 million gallons of slurry were released. Approximately 60 miles of the nearby Big Sandy River were contaminated.[14] Local water supplies were contaminated for days after the breech. Evidence for the contamination persisted for a decade after the release.[15]

Although the Buffalo Creek and Martin County disasters are perhaps the most graphic examples of impoundment failures, there are others that are less spectacular or are hidden from view. In 2010, the Wheeling West Virginia Jesuit University maintained a website that listed 65 impoundment failures that had released 748 million gallons, beginning with the Buffalo Creek disaster.[16] More recently, a 2013 report in the Washington Post, based on data from the Mine Safety and Health Administration, stated that there were 596 coal slurry impoundments in 21 states, of which 114 were in West Virginia.[17] The article further stated that tests of impoundment walls revealed flaws at all seven locations tested. Only 16 field tests out of a total of 73 met standards. These data suggest that more coal waste impoundment failures may occur in the future.

In the aftermath of the Martin County release, Congress asked the National Academies of Science to evaluate coal waste disposal practices.[18] This report, published in 2002, made numerous recommendations concerning the design of impoundments and suggestions for better ways to create, handle and dispose of this waste. The committee concluded that this was a responsibility that was shared by industry and government.

3.3   Coal Shipment

The pollution generated by the transport of coal from mines to power plants must be included as part of the unpriced consequences of coal combustion. Most of US domestic coal consumption (92.3%, as of 2015) is used for electricity generation and, of the coal delivered to coal-fueled power plants, approximately 70% was delivered by rail.[19] The diesel locomotives that haul coal emit particulate matter, nitrogen oxides, hydrocarbons and carbon monoxide.[20] The severity of that pollution was recognized by the US Environmental Protection Agency (EPA), which in March 2008 finalized a program to dramatically reduce emissions from diesel locomotives. The Agency estimated that, when fully implemented, these new protections would result in a reduction in particulate emissions by as much as 90% and NO_x emissions by as much as 80%, with "sizeable" reductions in emissions of hydrocarbons, carbon monoxide and other air toxics.[21] Without these controls, Agency scientists estimated that by 2030 locomotive and marine diesel engines would be expected to contribute more than 65% of the nation's mobile-source diesel fine particulate emissions in the US and 35% of national mobile source NO_x emissions. As is described in Section 4, these pollutants contribute to serious public health problems that include: premature mortality; aggravation of respiratory and cardiovascular disease; and exacerbation of existing asthma, acute respiratory symptoms and chronic bronchitis. Diesel exhaust has also been classified by the EPA as a likely human carcinogen.

The value of avoiding these emissions is significant. The EPA projected its emission control program for locomotives would, by 2030, generate monetized health benefits of $9.2 billion to $11 billion *per* year, assuming a 3% discount rate, or between $8.4 billion and $10 billion assuming a 7% discount rate. It compared that to the estimated annual social cost of the program in 2030, which it estimated to be $740 million. Thus the overall benefits of the emissions control program were estimated to outweigh social costs by between 9 : 1 and 15 : 1, depending on the discount tool applied.[20]

4 Combustion

4.1 A Brief History of Coal Combustion and Health

The pollutants produced by burning coal cause the most pervasive and varied adverse health effects associated with this energy source. Here it is important to reiterate the fact that "coal" is not a single entity. The carbon in coal is what makes it valuable as a fuel. This is what produces the heat and the carbon dioxide when it is burned. However, the makeup of various forms of coal, that is the number and concentration of non-carbon components, varies enormously from location to location. But, it is possible to make some general statements. Typically, the pollutants produced during combustion vary somewhat by the rank of the coal, with anthracite at the top, which is the best, followed by bituminous, sub-bituminous, and ending with peat. This is also the order for burial time and the application of heat and pressure that transform vegetation into coal over geological timeframes.

Burning coal was acknowledged as a health threat even before the industrial revolution, as seen from the title of John Evelyn's 1661 treatise on coal and its pollutants, "The Inconvenience of the Aer and Smoak of London, Together with some Remedies Humbly Proposed by J.E. Esq. To His Sacred Majestie and to the Parliament now Assembled." Inconvenience is hardly the word to be used today.

Several sentinel events mark the beginning of the era when it became evident that burning coal produced pollution that was immediately harmful and even fatal. The Donora, Pennsylvania, experience is central to the transformation that saw smokestacks as a sign of industrial might turn into an unacceptable sign of intolerable pollution.[22] In October 1948, almost half of Donora's 14 000 residents were sickened when atmospheric conditions trapped oxides of sulfur and other toxicants emitted by the factory that smelted iron and produced zinc-coated products. The chemical smog was so thick and the visibility was so poor that residents had to feel their way down the street by touching the buildings. Twenty people died and 400 were hospitalized. A similar, even worse event occurred in London in 1952 when the infamous killer fog paralyzed the city for days, killing around 12 000. The Donora Smog Museum, located in what was a Chinese restaurant, stands as a monument to the disaster and source of information about the disaster. Donorans proudly and correctly believe that the analysis of their catastrophe is what led to the cleaner air we take for granted today.

Although air in the US is demonstrably cleaner and healthier than it was in the past, coal smoke continues to be a huge threat to health elsewhere, particularly in China, where air quality is a serious problem. Indoor air pollution continues to be a major threat to health in developing nations because of the use of coal and other fuels for domestic energy production.[23]

An analysis of the Donora and London events triggered the steady progress and adoption of measures designed to control the deadly emissions of coal-fired power plants. The US result was the Clean Air Act (1963), a measure that was amended several times, which persists to this date and is arguably the most important public health measure of the 20th century.

As a result of a congressional charge, the US Environmental Protection Agency (EPA) began a study of pollutants that are emitted by coal-fired power plants.[24] This was an ambitious undertaking. In the words of the charge, the Agency was asked to "... perform a study of the hazards to public health reasonably anticipated to occur as a result of emissions by electricity utility steam generating units ..." To this end, the Agency measured the emissions from 684 units. Fifty-two of these were selected for a more intense scrutiny as they were considered to be representative of the group as a whole. The investigators identified 67 pollutants that were selected for further evaluation. Those with the greatest potential threat and whose primary route of exposure was inhalation were selected for the closest scrutiny. The list was augmented by toxicants whose route of exposure was non-inhalational if they were emitted in large quantities and were persistent and tended to bioaccumulate. Some of the toxicants, such as arsenic, lead and mercury, are probably familiar. Others such as acrolein, beryllium and cadmium are less familiar to most individuals. A summary of the results is shown in Table 1. Also, under the authority of the Clean Air Act, but unrelated to the study of power plants, the EPA identified six common pollutants that are known to affect human health and cause damage to property. These are known as the "criteria pollutants": carbon monoxide, lead, particulate matter, ozone, and oxides of nitrogen and of sulfur. The Agency is mandated to establish ambient air quality standards for each of these pollutants that are necessary "to protect public health, including the health of at-risk populations, with an

Table 1 Nationwide emissions of priority Hazardous Air Pollutants (HAP) identified in the study of hazardous pollutant emissions from electric utility steam generating units. (Reproduced from Lockwood.)[16]

HAP	1994 Emissions (tons per year)	Cancer risk^a	Toxicity^b
Arsenic (As)	56	3×10^{-6}	Long-term ingestion of small amounts may affect skin (hyperpigmentation, corns, and warts) damage peripheral nerves (painful sensation of "pins and needles") and increase risk of cancer of urinary bladder and lung
Beryllium (Be)	7.9	3×10^{-7}	In comparison with other elements (lead, chromium) Be exposure is insignificant. Most ingested Be is eliminated in the feces. Inhaled Be is more persistent. Inhalation of Be compounds (greater than 1 $mg\,m^{-3}$) may cause acute or chronic lung disease. The average Be concentration in urban air in the US is 0.2 $ng\,m^{-3}$, 1 ng = 1 billionth of a gram.
Cadmium (Cd)	3.2	2×10^{-7}	Cd accumulates in shellfish (observe fishing advisories), organ meats, lettuce, spinach, potatoes, grains, peanuts, soybeans, sunflower seeds and tobacco. Inhalation low levels of Cd for years or consumption of food with elevated Cd may cause kidney disease or fragile bones.
Chromium (Cr)	62	2×10^{-6}	Cr is a known carcinogen. Concentrations in air are typically less than 2% of those that cause respiratory problems in Cr workers. Avoiding tobacco smoke and older pressure-treated lumber minimize exposure.
Lead (Pb)	62	N/A	Children are more vulnerable that adults. Neurological problems include encephalopathy (global brain dysfunction) producing behavioral and cognitive deficits, damage to peripheral nerves, anemia, and kidney damage.
Manganese (Mn)	168	N/A	Major exposure comes *via* consumption of large amounts of grains, beans, nuts, tea, and nutritional supplements. Small amounts are inhaled. Accumulation in the brain, particularly in patients with liver disease, may cause symptoms similar to Parkinson's Disease.
Mercury (Hg)	51	N/A	Neurotoxin that affects brain development causing mental retardation. In adults, adverse effects may involve digestive system, lungs, kidneys, and immune systems. May be fatal.

Table 1 Continued.

HAP	1994 Emissions (tons per year)	Cancer risk[a]	Toxicity[b]
Nickel (Ni)	52	4×10^{-7}	There are multiple routes of exposure including inhalation. Food is the major source. The average concentration in air is 2.2 $ng\,m^{-3}$ and is attached to small particles. Concentrations found to cause cancer were 100 000 to 1 million times greater than that commonly in air. 10–20% if people are allergic to Ni.
Hydrogen chloride (HCl)	134 000	N/A	HCl removed from atmosphere in rain, limiting exposure from HCl released into the air.
Hydrogen fluoride (HF)	23 000	N/A	HF removed from atmosphere in rain, limiting exposure from HF released into the air, low concentrations of fluorine harden teeth and bones.
Acrolein	27	N/A	Inhalation causes irritation of nasal mucosa or other parts of the respiratory tract. Outdoor air concentrations range between 0.5 and 3.2 ppb. Minimum risk levels for chronic duration inhalation are not available and are about 3 ppb for exposures of less than 14 days. Environmental tobacco smoke is the major cause of exposure.
Dioxins[c]	0.00020	5×10^{-8}	Dioxins are probably carcinogens. They may cause a variety of skin problems, including chloracne. Type II diabetes and other endocrine disorders have been attributed to dioxins.
Formaldehyde	29		Formaldehyde decomposes to formic acid and carbon monoxide within a day. Air concentrations in the highest areas are 10–20 ppb. Many home products release formaldehyde and indoor air concentrations are usually higher than in outdoor air. Formaldehyde is an irritant and is dangerous to life at a concentration of 20 ppm. It is likely to be a carcinogen.

[a]Cancer Risk = highest cancer risk for maximally exposed individual due to inhalation of the HAP for 70 years at the highest presumed HAP concentration. (For details of modelling see ref. 16, Section 6.1.1 and the health risks sections of the appendices in volume 2.)[16]

[b]Toxicity information was obtained from the Agency for Toxic Substances and Disease Registry ToxGuides™. (Available at: http://www.atsdr.cdc.gov/toxguides/index.asp, Public Health Statements, http://www.atsdr.cdc.gov/PHS/Index.asp, or ToxFAQs™, http://www.atsdr.cdc.gov/toxfaqs/index.asp; accessed May 26, 2010.)

[c]Dioxin emissions are the summation of dioxin equivalents for each member of this family relative to 2,3,7,8-tetrachlorodibenzo-p-dioxin.

adequate margin of safety." These standards are known as National Ambient Air Quality Standards or NAAQS and are revised periodically as science advances and more is learned about the health effects. This rule-making process is complex and includes a thorough review of the literature, advice from a science advisory committee and the publication of a proposed rule. After the Agency receives comments from the public and other stakeholders, a final rule is published in the Federal Register. Politics and legal battles are typical features of the rule-making process. The NAAQS as of January 1, 2017 are shown in Table 2.

4.2 Particulate Matter

Increasing evidence suggests that particulate matter may be the most important of all the coal-related criteria pollutants in its adverse impacts on health. The more that is learned about it, the more it becomes evident that strict controls are needed to provide the adequate margin of safety called for by the Clean Air Act.

Particulate matter (PM) is not a single entity. Technically, it is an atmospheric dispersion of solids and liquids: an aerosol. Advances in science have led to changes in how PM is classified in the NAAQS. Initial methods, crude by today's standards, lumped virtually all PM into a single category. Improved monitoring methods led to an evolution in PM classification. PM size is now the mainstay. For regulatory purposes and in most epidemiological studies, PM is described in terms of its aerodynamic diameter. This is a somewhat confusing term, since PM exists in many shapes. Particles with the same aerodynamic diameter settle to the ground at the same rate under the influence of gravity regardless of their actual size and shape. The aerodynamic diameter of PM of interest ranges between 100 and 0.01 μm. $PM_{2.5}$ refers to particulates 2.5 μm or less in aerodynamic diameter and PM_{10} refers to particulates 10 μm in aerodynamic diameter. Ironically, the smallest particles are the most threatening to health. This is because they travel into depths of the lungs, to the alveoli where the exchange of carbon dioxide for oxygen takes place. It is here where the complex inflammatory and immunological processes begin. Larger particles fail to reach the alveoli as they are trapped by nasal hairs, mucous membranes of the trachea, bronchi and other portions of the pulmonary system, and usually cause much less harm.

PM is further classified by the mechanism of formation. Primary PM is formed directly in the combustion chambers of power plants and internal combustion engines (especially diesel engines). Wildfires, roads and construction sites are also among the important sources of primary PM. Large amounts of the more abundant secondary PM are formed by reactions among chemicals in the atmosphere. Oxides of sulfur and nitrogen are of central importance in these reactions that form secondary PM.

When the EPA first began to monitor PM, methods were crude by the standards of today. PM measuring techniques only allowed for reports of

Table 2 National Ambient Air Quality Standards as of January 1, 2017 (National Ambient Air Quality Standards).[a]

Pollutant	Primary/Secondary	Averaging time	Level	Form
Carbon monoxide	Primary	8 h 1 h	9 ppm 35 ppm	Not to be exceeded more than once per year
Lead	Both	Rolling 3 month average	0.15 $\mu g\,m^{-3}$	Not to be exceeded
Nitrogen dioxide	Primary	1 h	100 ppb	98th percentile of one-hour daily maximum concentration averaged over 3 years
	Primary and secondary	1 year	53 ppb	Annual mean
Ozone	Primary and secondary	8 h	0.070 ppm	Annual fourth-highest daily maximum 8-h concentration averaged over 3 years
Particles PM$_{2.5}$	Primary	1 year	12 $\mu g\,m^{-3}$	Annual mean averaged over 3 years
	Secondary	1 year	15 $\mu g\,m^{-3}$	
	Primary and secondary	24 h	35 $\mu g\,m^{-3}$	98th percentile averaged over 3 years
Particles PM$_{10}$	Primary and secondary	24 h	150 $\mu g\,m^{-3}$	Not to be exceeded more than once per year over 3 years
Sulfur dioxide	Primary	1 h	75 ppb	1 h daily maximum concentration averaged over 3 years
	Secondary	3 h	0.5 ppm	Not to be exceeded more than once per year

[a]Source: Environmental Protection Agency, https://www.epa.gov/criteria-air-pollutants/naaqs-table, accessed January, 2017.

PM_{10}. The frequency of reporting was limited severely. Now measurements of $PM_{2.5}$ are reported from scores of sites, often on an hourly basis. These results are in the public domain and are used to report and forecast air quality at multiple sites across the US on the EPA Enviroflash website.[25] These reports allow citizens to take air quality into account, particularly when planning strenuous outdoor activities or when activities involve sensitive populations such as children, asthmatics and individuals with cardiopulmonary diseases. As statistical and epidemiological methods have become more sophisticated, it has become possible to merge air quality data with health outcomes to provide greater insight into the health effects of pollution.

The most recent advances in the science of merging PM data with health data have shown close linkages between timed events, such as an acute stroke, and a brief spike in the $PM_{2.5}$ concentration.[26] The importance of PM size was established nearly 20 years ago.[27] In that study, the investigators found that increases in mortality were significantly associated with increasing atmospheric concentrations of PM_{10}, $PM_{2.5}$, and sulfate particle concentration. The association was strongest for the smallest particles. This is the size of particles commonly formed by combustion. This and other studies validate the EPA's regulatory emphasis on these particles.

The next step on the horizon may involve the characterization of the chemical constituents of PM and relating these to disease. An important step in this direction came in 2000 when an elemental analysis of PM was performed in an attempt to identify and characterize sources of the pollutant.[28] The investigators performed elemental analyses of $PM_{2.5}$ obtained from six US cities and found measurable amounts of silicon, aluminum, calcium, iron, manganese, potassium, lead, bromine, copper, zinc, sulfur, selenium, vanadium, nickel and chlorine. They subjected their results to a statistical technique known as factor analysis and were able to identify selenium as the element in the analysis that characterized emissions from coal-fired plants. Silicon and lead were associated with PM derived from the earth's crust and motor vehicles, respectively. Both the silicon and selenium factors were associated with increases in the daily mortality rate. For a $10 \ \mu g \, m^{-3}$ increase in the concentration of $PM_{2.5}$, there was an increase of 3.4% in the daily mortality associated with lead, the mobile source factor. For the coal factor, the increase in mortality for an identical concentration was 1.1%. Both were statistically significant. This was a landmark study and has been cited by others more than 1000 times.

One of the hallmarks of valid science is the ability to replicate a result. In a subsequent investigation, an international group from the US and Canada evaluated the relationship between $PM_{2.5}$ and ischemic heart disease (IHD).[29] This group compiled vital status and cause-of-death data from almost 450 000 adults in 100 US metropolitan areas for the years 1982 to 2004. They focused exclusively on IHD as an endpoint. They characterized $PM_{2.5}$ based on analyses of chemical constituents and source-associations. They found that the IHD risk associated with exposure to coal-derived $PM_{2.5}$

was approximately five times greater than exposure to $PM_{2.5}$ in general (*i.e.* when not broken down by source). $PM_{2.5}$ associated with diesel traffic was also associated with an elevated risk for developing IHD. Their conclusion was predictable: the largest IHD health benefits may be achieved by reducing "fossil fuel combustion exposure, especially from coal-burning plants." $PM_{2.5}$ exposures from wind-blown soil and biomass combustion were not associated with IHD mortality.

The mechanisms by which $PM_{2.5}$ causes disease are complex, as illustrated in Figure 1. These mechanisms were described in more detail in an earlier publication.[16] Briefly, the process begins with inhalation of the particles. Thus, the lung is central to these processes. In the alveoli the particles produce oxidative stress and inflammation. These reactions, combined with systemic effects, cause constriction of blood vessels and affect the cells that

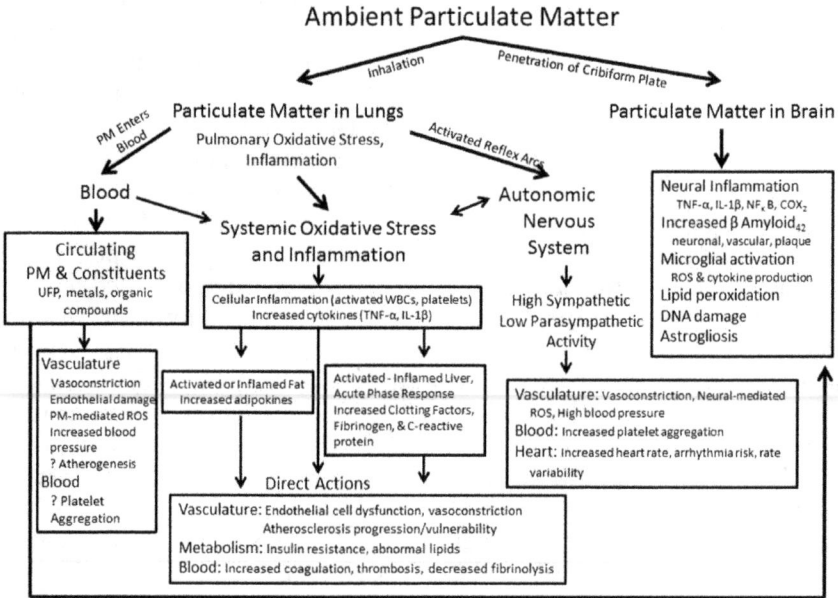

Figure 1   Pathways linking cardiovascular and nervous system disease to particulate exposure. Particulates cause acute, subacute or chronic effects after they enter the blood, produce systemic oxidative stress and inflammation, and affect the autonomic nervous system, and enter the brain *via* the cribiform plate or the cerebral circulation. (Abbreviations: UFP, ultrafine particles; TNF-α, tumor necrosis factor alpha; IL-1β, interleukin 1-beta; WBC, white blood cell; ROS, reactive oxygen species; CRP, C-reactive protein; NFκ B, nuclear factor kappa-light chain enhancer of activated B cells; COX2, second isoenzyme of cyclooxygenase, β-Amyloid42, a 42 aminoacid protein associated with Alzheimer's disease; IL-6, interleukin-6; TNF-α, tumor necrosis factor alpha; – all are molecules mediating or responding to oxidative stress and/or inflammation).

(Reproduced with permission from Lockwood, MIT Press.[16])

line blood vessels (endothelial cells), leading to, or contributing to, atherosclerosis, which is in turn a major risk factor for ischemic heart disease, stroke and hypertension. These problems are compounded by increases in the tendency for blood to clot, a mechanism that may cause a heart attack or a stroke. Metabolic effects are thought to include insulin resistance, a factor associated with Type II diabetes mellitus. The lungs are innervated by the autonomic nervous system. PM activates reflexes in this part of the nervous system, contributing to the constriction of blood vessels and elevations in blood pressure. To summarize, $PM_{2.5}$ exposure *via* the lungs triggers pathophysiological processes that underlie heart disease, cancer, chronic lower respiratory diseases (such as emphysema and chronic bronchitis, components of chronic obstructive pulmonary disease or COPD, and asthma) and stroke.

Fine PM exposure has also been linked to neurodegenerative disorders including Alzheimer's disease and Parkinson's disease, two diseases that are particularly common among the elderly.[30,31] In the case of diseases of the central nervous system, the route of exposure is partially *via* direct entry into the olfactory system. The smallest particles enter the neurons mediating smell directly, and pass through small holes in the bone at the base of the skull in the frontal region (the cribiform plate). These neurons pass to the olfactory bulbs, a part of the limbic system of the brain that mediates many functions. These include memory, a function of the hippocampus, which is a critical part of the limbic system. In the brain, PM activates neural enzymes that mediate inflammation and are linked to the production of β-amyloid, a protein that is characteristic of Alzheimer's disease.[29]

It is difficult to predict what effects climate change might have on PM.[32] It is likely that trends in fuel choices, such as the transition from coal to natural gas, will have a greater effect on the atmospheric PM concentration than will the climate itself.

4.3 Ozone

Ozone is a pale blue gas. It has a pungent odor, somewhat similar to that of chlorine gas. It is formed by natural processes, *e.g.* lightning, and as the result of human activity. It is present in the stratosphere, where it protects us, plants, and other life forms from the harmful ultraviolet rays of the sun. It is also present in the troposphere, the lowest layer of the atmosphere, where it is a major constituent of ground-level smog. It is the tropospheric ozone that affects health and is regulated by the EPA.

Ozone is one of the most powerful of all naturally-occurring oxidizing agents. It "attacks" a large variety of molecules, living organisms, and their tissues. This is why it is of importance in causing or exacerbating a variety of diseases. It is a potent irritant in the lungs and is believed to cause or exacerbate a variety of pulmonary diseases such as asthma and chronic obstructive pulmonary disease (emphysema and chronic bronchitis). For this reason it, along with $PM_{2.5}$, is a principal determinant of air quality, as reported online and in many newspaper weather reports.

Ground-level ozone is formed by complex chemical reactions involving oxides of nitrogen and other organic compounds such as carbon monoxide, volatile organic compounds (VOCs) and methane.[31] The energy needed to drive these reactions comes from sunlight. Some of the VOCs occur naturally, particularly terpenes, which are VOCs produced by plants. Others are the result of human activity and arise from gasoline vapors, paints and virtually any imaginable volatile compound. Tropospheric ozone is a greenhouse gas and has made a significant contribution to climate change. But, things are not all that simple. For example, since ozone is a powerful oxidizing agent, it scavenges other greenhouse gases from the atmosphere, particularly methane, which is also a powerful greenhouse gas. Large amounts of methane are released by natural gas drilling operations, compressor stations and pipelines, particularly in metropolitan areas where old natural-gas distribution pipes are often leaky (see Lockwood).[32] Thus, quantifying the respective roles of the chemical agents that underlie ozone production is complex.

The current 8 hour EPA standard for ozone is 70 parts *per* billion (ppb). It was recently lowered to this level from 75 ppb after a long and contentious process in which science and environmentalists clashed with industry.[16,31] When the EPA lowered the standard to 75 ppb during the George W. Bush administration, it acted against the unanimous advice of the Clean Air Scientific Advisory Committee (CASAC), which advised setting the standard to between 60 and 70 ppb. The EPA actions were widely attributed to political considerations. The standard was revisited between 2009 and 2010 during the Obama administration. After the usual rule-making steps of publication in the Federal Register and the receipt of numerous comments from stakeholders, the Agency announced that it was suspending its review. Again, this was thought to be a political decision designed to deflect criticism directed toward the administration during a pre-election period. As one might expect, lawsuits ensued and the Agency ultimately entered into a consent decree that established an October 15, 2015, deadline for finalizing the ozone rule. Industry representatives called for no change, while health and environmental advocates called for a standard of no higher than 60 ppb or at least at a level that was between 60 and 70 ppb. Advocates for the 60 ppb standard argued, unsuccessfully, that this was warranted in view of numerous reports in which normal individuals developed decrements in measurements of lung function when breathing ozone at a concentration of 60 ppb under laboratory conditions. Health advocates argued that since these so-called normal volunteers were in many cases trained athletes, a standard higher than 60 ppb would fail to provide the adequate margin of safety for sensitive populations called for by the Clean Air Act. Industry representatives argued that a new standard was not needed and would be harmful to their interests. Further legal action is likely.

The very short lifetime of tropospheric ozone makes it likely that there will be a substantial amount of variation in its concentration among different locations. Thus, local factors will be major determinants of ozone concentrations in the future.[16] Increases in temperature will favor ozone production, particularly in cities where emissions of nitrogen oxides are high due to motor vehicles and the heat island effect that makes cities warmer than the surrounding region. Predictions about precipitation in the future suggest that humidity will be high in the southeastern US and low in the southwest portion of the nation. Since water vapor reacts with ozone, ozone concentrations are more likely to rise in areas with low humidity and *vice versa*. Increased lightning and isoprene concentrations due to more severe thunderstorms and the fact that carbon dioxide stimulates the growth of some plants, and thus the release of isoprene, will add to the complexity of predicting future ozone concentrations. In a detailed predication of the effects of climate change on 50 cities in the US, the authors predicted that, on average, the one-hour maximum would increase by 4.8 ppb.[33] The largest increase was expected to be 9.6 ppb. They further predicted that there would be a 68% increase in the number of days when the then-standard of 75 ppb would be exceeded.

These and other predictions suggest that the EPA and equivalent regulatory bodies in other countries should anticipate these increases and set lower permissible concentration limits that anticipate the rise. This would provide the best protection for these anticipated increases. This has been referred to as a "climate change penalty".[34]

4.4 Oxides of Nitrogen and Oxides of Sulfur

These two criteria pollutants are both irritants and exacerbate respiratory diseases. As noted above, oxides of nitrogen are involved in the production of ground level ozone or smog. Both of these chemicals are highly reactive and combine with other atmospheric constituents to form secondary PM. Both are acidic. Oxides of nitrogen are formed by the chemical combination of atmospheric nitrogen and oxygen in combustion chambers of boilers and internal combustion engines. Oxides of sulfur are formed from the sulfur in coal when it is burned. The sulfur content of coal varies substantially. Coal from the Powder River Basin, in Montana and Wyoming, has a relatively low sulfur content and is desired because of this characteristic. Some coal mined in China has an extremely high sulfur content. In a move designed to improve the air quality in China, the Chinese government banned the importation of high-sulfur coals.[35]

In 1990 the Clean Air Act was amended. This included the so-called Acid Rain Amendment, designed to curb the emission of oxides of nitrogen that were contributing to urban air pollution and acidifying lakes, particularly those in the northeastern portion of the US. In some lakes the degree of acidification was so severe that they were unable to support the species that had been typical of those bodies of water. Although it is hard to imagine in this day of high partisanship, not only was the amendment proposed by a Republican president, but it was passed by large bipartisan majorities on both the House of Representatives and the Senate (401–21 and 89–11,

**Sulfur Dioxide and Nitrogen Oxides Emissions
Under Clean Air Act, Acid Rain Program**

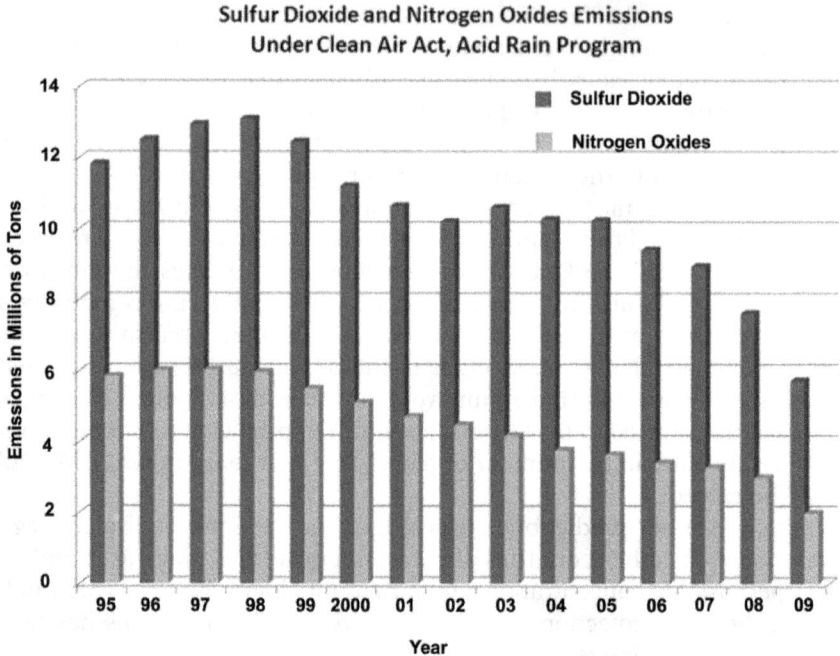

Figure 2   US SO_x and NO_x emissions from 1995 to 2009, showing decreases attributable to the Environmental Protection Agency's Acid Rain Program of the Clean Air Act.
(Reproduced with permission from Lockwood, MIT Press.[16])

respectively). As a result, there have been dramatic reductions in the emissions of these two pollutants, as shown in Figure 2, with corresponding dramatic reductions in the adverse health impacts caused by these two pollutants.

4.5   Mercury

Although mercury is not a criteria pollutant, it is a dangerous pollutant and is emitted in substantial quantities by coal-fired power plants, as shown in Figure 3.[36] Elemental mercury is a shiny silver liquid at room temperature. When small quantities are placed on a flat surface, it tends to form balls that move easily when disturbed. These properties make spills difficult to contain and remediate. Most mercury is in some chemically combined form of the element.

Important anthropogenic sources of mercury emissions are shown in Figure 3. These include gold mining and smelting, and burning coal. The emissions associated with artisanal gold production are particularly difficult to estimate due to the fact that this practice is largely unregulated and widely distributed.[35] Examinations of ice cores obtained in Wyoming shows that

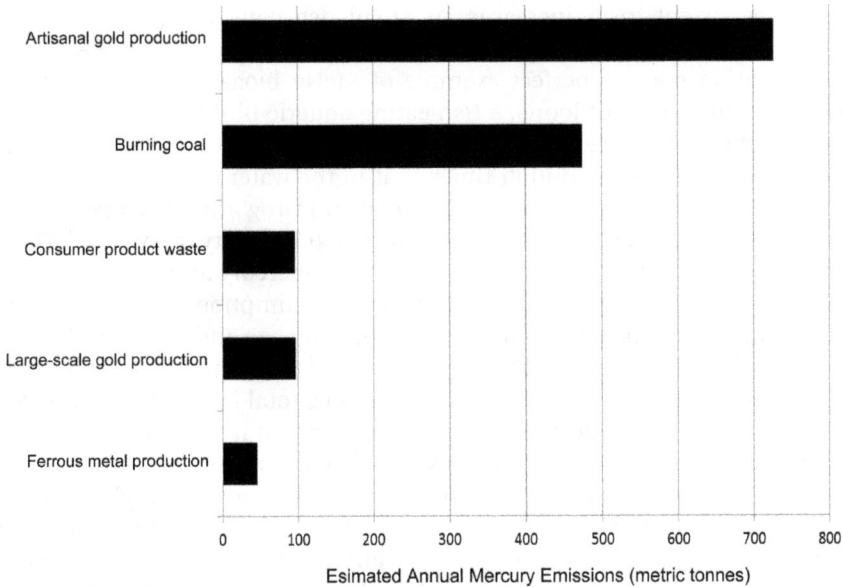

Figure 3 Anthropogenic mercury sources, world-wide. Data units are in metric tonnes *per* year (one metric tonne equals 2200 lb).
(Data from United Nations Environmental Program.[36])

the deposition of mercury has increased steadily during the industrial era, most likely due to burning coal, with blips due to gold production. Periodic peaks that occur in this record are timed to major volcanic eruptions, such as Mount Saint Helens, Washington in 1980 and Tambora, Indonesia in 1815,[34] as volcanoes and geothermal sites are natural sources of mercury.[34] One of these peaks is associated with the California gold rush. Mercury that is deposited in forests, oceans and other locations may be re-emitted as the consequence of forest fires and wave action. This creates a cycle wherein mercury moves from one location to another.

Mercury, regardless of its chemical form, is released into flue gases when coal is burned. Although much of the mercury thus released falls to the ground close to a point source, some enters the upper atmosphere and may travel long distances.[34] For this reason, mercury emitted by Chinese power stations contaminates waterways in Alaska.

Atmospheric mercury is contained in small particles that may form the nidus for the condensation of water into droplets of rain. This contaminated rain water eventually finds its way into rivers, streams and the oceans. In the silt and dirt-laden bottoms of waterways, bacteria convert the mercury into methylmercury (MeHg). Methylmercury is the chemical form of the element that poses the greatest threat to health. MeHg enters the food chain, as it easily crosses biological membranes, such as the lining of the gastro-intestinal tract and the placenta. Once in an organism, it stays there or bioaccumulates; since it is not broken down or excreted, the concentration of this persistent toxin increases. As small fish consume MeHg-laced organisms and are eaten by ever-more voracious predators, the MeHg concentration rises. A perfect example of MeHg bioaccumulation became evident from a study of loons, a fish-eating aquatic bird that is common in lakes in the Northeastern US.[37] The study showed that the MeHg concentration may rise to one million times that in the water.

The EPA, in a collaboration with the Food and Drug Administration (FDA), publishes a national listing of fish advisories (NLFA) every two years.[38] These advisories use data about the concentration of mercury and other toxins in waterways and provide evidence-based fish consumption advice to vulnerable populations, such as women of childbearing age and children. In 2011 the states issued a total of 223 new fish advisories, raising the total number to 4821 advisories, covering 42% of the nation's total lake acreage and 36% of the nation's total river miles. Most fish consumed in the US is purchased in grocery stores or fish markets. Many display information about their products, including data about sustainability. Fish to avoid include shark, king mackerel, tilefish, marlin and orange roughy. These are at the top of the food chain and are the most likely to contain the largest amount of mercury. Cooking does not reduce the mercury content of fish. A fact sheet entitled "Healthy Fish, Healthy Families" is available from Physicians for Social Responsibility.[39]

Mercury damages the central and peripheral nervous systems; that is, the brain, spinal cord, and the nerves that enter and exit these structures. The World Health Organization lists the following symptoms of mercury toxicity: tremors, insomnia, memory loss, headache and, importantly, cognitive and motor dysfunction.[40] Mercury may also affect the kidneys and cause kidney failure. Mild, subclinical effects that involve the central nervous system or kidneys may be present.

Unfortunately, aside from avoiding fish that contain the largest amount of mercury, there is little that individuals can do to avoid mercury by modifying behavior. However, individuals and stakeholder groups can exert pressure on officials demanding reductions in the burning of coal, and the phase-out of products that contain mercury such as some switches, thermometers and medical devices. Fluorescent light bulbs contain mercury and must be disposed of properly.

5 Costs of Burning Coal

Electricity is essential to modern economies. This is why developing nations place a heavy emphasis on generating electricity at the lowest possible cost. China and India are prime examples. Large numbers of coal-fired electricity generating units have been built to fulfill this need. All too often the cost of electricity has been portrayed as the cost charged for the amount of electricity consumed, in kilowatt hours. This is what is presented, for example, on your electrical bill and is commonly known as the price. The full costs of electricity would encompass, for example, the costs attributable to the effects of the associated release of pollutants. These costs are referred to as the hidden costs of energy or externalities, or the unpriced cost of electricity, and have been largely ignored until recently. A systematic approach to quantifying these costs was undertaken in the US by the National Research Council.[41] Under the Energy Policy Act of 2005 (Public Law 109-58), the Council was required "to define and evaluate the health, environmental, security, and infrastructure costs and benefits associated with the production and consumption of energy that are not or may not be fully incorporated into the market price of such energy." The report concluded that the health damages that were attributable to the 406 coal-fired plants that formed the basis of the study came to $62 billion *per* year in 2007 dollars. This averaged out to $156 million *per* plant. The largest, oldest and most damaging of these electrical generating units were located along the Ohio River and the Appalachian region of the US.

A more recent and more comprehensive analysis of the externalities associated with the entire life-cycle of coal, including mountain-top removal mining practices and the waste stream generated by combustion, found that the full cost of coal was likely to be between one third to one half billion dollars *per* year.[42] This study concluded that including these externalities in the price of electricity would double or triple the cost *per* kilowatt hour paid at the meter. Since that report was published, the cost of electricity generated from solar and wind power has fallen dramatically, making electricity generated by these renewable sources more than competitive with coal. Once installed, these sources of renewable energy do not have any fuel costs: sunlight and wind are free.

As a part of the Clean Air Act, the EPA is required to make periodic reports of the Act's costs and benefits of the Act to Congress. The most recent of these was published in 2010 and covers the interval between 1990 and 2020.[43] Although this report includes all of the pollutants regulated by the Agency, a substantial portion are the result of burning coal. The Agency projects that by 2020 the total health benefits of the Clean Air Act will be in the order of two trillion dollars *per* year, at a cost to industry of around 65 billion dollars. About $10 billion of the cost is attributed to rules mandating better pollution controls on power plants. Many of the benefits listed by the Agency are as one might expect: reductions in hospitalizations, fewer deaths (monetized by lost years of life), emergency room visits and other medical expenses. They also include reductions in lost days at work, since productivity is affected, and missed days of school, since reimbursements to many schools depend on attendance. Arguably, the Clean Air Act, which also removed lead from gasoline, might be the most important legislation affecting public health in the last century. Even more health benefits will be accrued if the Agency's Clean Power Plan, put forth by the Obama administration, takes full effect. The fate of the Clean Air Act and regulations promulgated under that authority is uncertain at this time (early 2017), given the beginning of the Trump administration and their vows to eliminate rules and regulations.

6 Post-combustion Waste: Coal Ash

The harm that coal can inflict on human health does not end when the coal is burned. Rather, the waste product that remains after combustion is highly toxic and poses multiple threats to human health. This waste product is known in the industry as coal combustion waste and, more commonly, as coal ash. Since coal ash constitutes a massive and toxic waste stream, which poses threats to human health that differ in some respects from those of coal combustion pollutants, and since coal ash follows different pathways through the environment, we are affording it separate and detailed examination.

6.1 Coal Ash Defined

The term "coal ash" is used to encompass several different forms of coal combustion waste. "Fly ash" consists of the fine powdery particles of minerals, plus a small amount of carbon, that is carried up the smokestack by exhaust gases. "Bottom ash" is a coarser material that falls to the bottom of the furnace. "Boiler slag" is created from the molten bottom ash that, when cooled in contact with water, forms pellets of a hard, glass-like material. Flue gas desulfurization (FGD) waste is the byproduct of air pollution control systems used to reduce sulfur dioxide emissions from coal-fired power plants. Scrubbers are used to spray lime or limestone slurry into the flue gas, where it reacts with the sulfur to form calcium sulfite. The calcium sulfite is processed to make synthetic gypsum; what remains is FGD waste. Fluidized bed combustion (FBC) wastes are generated by a specialized combustion technology in which a heated bed of sand-like material is suspended (fluidized) in a rising jet of air. FBC waste may include fly ash and bottom ash and tends to be more alkaline because of the limestone used in the process.

Four factors contribute to the dangers that coal ash poses to health. Three are inherent: the ash's toxicity; its volume; and its characteristics of persistence and mobility in the environment. The fourth is a factor more subject to human control: the long-term inadequacy of the most common disposal methods.

6.2 Coal Ash Toxicity

When coal is burned to generate electricity in coal-fired power plants or in smaller industrial boilers, the non-combustible components remain as waste. This waste contains many toxic components, occurring in a far greater concentration than in the coal itself. While the specific toxic contents of coal ash may vary somewhat, depending on where the coal was mined, coal ash commonly contains some of the world's most deadly pollutants, including toxic metals that can cause cancer and neurological harm to humans. These include arsenic, which is known to cause cancers of the lung, kidney, bladder, skin and other organs; hexavalent chromium, which can cause stomach cancer; and lead and mercury, both of which can permanently damage the developing brain. In addition, coal ash is likely to contain a host of other metals, all of which can be toxic. In fact, the list of dangerous metals commonly occurring in coal ash can be said to run from A to Z, as it includes: aluminum, antimony, barium, beryllium, boron, cadmium, chlorine, cobalt, manganese, molybdenum, nickel, selenium, thallium, vanadium and zinc.[44] The health effects of many of them are described in Table 1. Especially where there is prolonged exposure, these metals can cause several types of cancer, heart damage, lung disease, respiratory distress, kidney disease, reproductive problems, gastrointestinal illness, birth defects, impaired bone growth in children, nervous system impacts, cognitive deficits, developmental delays and behavioral problems. In short, coal ash toxics have the potential to injure all of the major bodily organ systems, damage physical health and development, and contribute to mortality.

The health effects described above are known to be associated with exposure to some of the dangerous metals in coal ash. Yet these are the effects as the toxicants act individually. Little is known about the potential for additive effects on human health when these substances intermix, as happens routinely in coal ash. Concurrent exposure to several contaminants raises the possibility of synergistic interactions and produces a greater chance of increased risk to health. Effects on the human body may be intensified where several coal ash contaminants share a common mechanism of toxicity, or where several substances affect the same body organ or system. For example, aluminum, manganese and lead all have adverse effects on the central nervous system; barium, cadmium and mercury all have adverse effects on the kidney. Yet despite the common co-occurrence of multiple contaminants in coal ash, the phenomenon of their interaction has been little studied in the laboratory. Such study would be complex, in part because the precise mixtures, concentrations and interactions among the multiple substances that occur in coal ash are difficult to predict under the haphazard conditions of coal ash disposal in the real world.

Apart from the phenomena of co-occurrence, several other factors influence and in fact intensify the toxicity of coal ash. The air pollution control technologies that are used to reduce air pollution from coal-fired power plants – scrubbers, selective catalytic reduction, and activated carbon injection technologies – capture an increasing proportion of the mercury and other hazardous air pollutants that would otherwise go out of the smokestacks. This is a highly beneficial advance, as far as air quality is concerned; however, the captured pollutants are often disposed of by adding them to coal ash, thus increasing the toxic burden of the ash. In other words, these laudable technological advances may serve to replace one environmental hazard with another. As the technologies used for smokestack capture of dangerous pollutants improve, coal ash is likely to grow increasingly toxic and dangerous.

Another factor that adds to the toxicity of coal ash is the practice employed at some power plants of mixing coal with other fuels and wastes, such as used tires and even hazardous wastes. Utilities that manage coal ash in ponds may also mix coal refuse in with the coal ash. Coal refuse is waste material that is the by-product of coal mining, screening or processing. It is a highly acidic waste, and the resulting mixture is significantly more toxic and more prone to release metals into the environment than is coal ash alone.[45] This practice greatly increases the cancer risk to nearby residents who draw their drinking water from wells.[43]

6.3   Coal Ash: A High-volume Waste

In the United States, coal ash constitutes the nation's second-largest industrial waste stream, second only to mining wastes. As of 2014, it was estimated that coal-fired power plants generated about 140 million tons of coal ash – fly ash, scrubber sludge and other combustion wastes – every year. This waste product is kept in over 1000 operating coal ash landfills and ponds, as well as several hundred inactive or "retired" coal ash disposal sites.[46]

Coal ash is a massive problem not only in regard to volume; it is also a major source of toxic pollutants, especially into waterways. In the US, the federal Environmental Protection Agency (US EPA) has acknowledged that coal-burning power plants are responsible for at least 50% to 60% of the toxic pollutants discharged into surface waters of the US by all industries.

6.4   Coal Ash Persistence and Mobility in the Environment

The concern over coal ash contamination reflects the difficulty of assuring that the large quantities of this dangerous industrial waste are kept – permanently – from contact with humans, our food and our drinking water. Coal ash components are elemental substances; as such, they persist in the environment. They may be contained by the disposal process or they may disperse, but they do not disintegrate or lose their toxicity; they never really "go away." They remain and, if not safely contained, may continue to pose exposure risks.

The difficulty of containment arises in part because toxins in coal ash are fairly mobile in the environment. The toxic materials in coal ash have the capacity to "leach" or dissolve in water and percolate through the earth and to contaminate surrounding bodies of water and drinking water supplies. Chemicals move at different rates through groundwater, so when contaminants leach out of coal ash, some take longer than others to reach those places where they may expose humans to risk. The US EPA has conducted sophisticated modeling to estimate how long leaching substances would take to reach their maximum concentrations in well water. When the dissolved coal ash toxics, called "leachate," is contained in unlined coal ash ponds (discussed in the following section), the median average time until peak well water concentrations would occur is estimated to be 74 years for selenium, 78 years for arsenic and 97 years for cobalt.[47] This indicates, among other things, that coal ash contaminants, once released into groundwater, can endanger health for decades or even centuries. It further indicates that testing for coal ash contaminants in water, conducted at any given time, may not give an adequate reading of potential risks.

The dissolved coal ash toxics can also enter surface waters, like rivers and streams, where they may endanger public health and the environment by contaminating water used for drinking supplies, or by contaminating fish that have been exposed to coal ash-exposed water or sediments. In addition to waterborne routes, coal ash can also go airborne, traveling through the air as fine particles, or may travel over the ground and other surfaces, due to erosion, runoff, or settling dust.

7   Disposal of Coal Ash

7.1   Common Disposal Techniques

Electricity generating plants typically dispose of coal ash in ponds, landfills or dumps. Surface impoundments, also known as coal ash ponds, contain ash mixed with water. These so-called "ponds" may be quite large – some have a surface area of more than 1000 acres – yet may be constructed only with earthen walls. Where the ash is not mixed with water, it can be disposed of in landfills. These are typically located on the grounds of the power station where the coal was burned, so as to avoid transport costs or, alternatively, the ash may be transported to an offsite landfill. Finally, some coal ash is disposed of in dumps or dropped into mines, active or abandoned, where the coal ash is utilized as fill. In 2007, an estimated 21% of US coal ash was disposed of in surface impoundments, 36% was disposed of in landfills, 5% was disposed in mines, and 38% was recycled in what is referred to as "beneficial reuse," discussed below.[48]

All these disposal systems are designed to avoid the various potential pathways to exposure. However, all have proven to be fallible. The danger that surface impoundments pose to communities and environments downstream is so great that the US EPA has established a system that rates those dangers. A "high" hazard rating indicates that a dam failure is likely to cause loss of human life; a "significant" hazard rating means that failure of the impoundment would cause significant economic loss, environmental damage, or damage to infrastructure. As of 2009, the US EPA found that of 431 ash ponds that had been rated, 50 – more than one in ten – had a "high" hazard rating, indicating a potential threat to life. An additional 71 had a "significant" rating.[49] The actual number of coal ash dams with high and significant hazard ratings was likely higher, given that almost 200 coal ash dams had not yet been rated at the time when EPA conducted its study.

Other disposal systems are not without their own dangers. In ponds, landfills and dumps, toxic substances not only leach into water but also spill, leak or are discharged into rivers, streams, lakes, or can contaminate underground aquifers (groundwater). And where ash ponds are allowed to dry, or where ash is stored in dry form, it can blow onto neighboring homes and gardens. Instances exist in which all of these pathways have resulted in measurable risk or damage to human health. Further discussion of these dangers is presented in Section 8.

7.2   Dilution as Disposal

Besides containment, another approach to risk reduction for coal ash is dilution. It is often said about toxic substances that "the dose makes the poison." Consequently, a convenient way to address their presence in the environment is to dilute their concentration, such as by mixing them with sufficiently large quantities of water, so that humans are exposed to quantities below recognized thresholds for harm. This approach to the handling of coal ash, however, encounters several shortcomings. Leaching may allow slow-moving plumes of toxics eventually to reach humans at dangerous levels of concentration; this is the danger that leaching from ash ponds poses to drinking-water wells. Coal ash toxics may also enter environments where a background degree of contamination already exists, thus creating cumulative levels that could exceed safe thresholds. This concern applies, for example, when arsenic from coal ash enters already-contaminated rivers, and when mercury from coal ash is added to the bioaccumulated toxic burden in already-contaminated fish. In addition, as the science of toxicology advances, maximum permissible concentrations once deemed safe for some toxic substances, for example arsenic, must be lowered, as further study reveals that harm to health occurs at levels previously considered to be safe. Finally, measurements simply are not made for some coal ash toxics, whether in drinking water or in other potential sources of exposure. Related concerns about drinking water are discussed below.

7.3   "Beneficial Reuse" Instead of Disposal

As an alternative to disposal, coal ash is reused in several industrial and construction applications. Disposal, after all, requires large tracts of available land and entails additional expenses, so the utilities and industries that generate coal ash may opt for recycling, which they call "beneficial reuse." Reuse offers significant economic benefits: companies generate income from its sale and avoid costs of its disposal.

Reuse takes a number of forms. Fly ash, which hardens when mixed with water and limestone, can be used as a substitute for Portland cement in making concrete. Bottom ash is sometimes used as an aggregate in road construction and concrete, and FGD gypsum sometimes substitutes for mined gypsum in agricultural soil amendments and in making wallboard. Ash is also used in structural fills in the building of overpasses and other road construction projects; spread as an anti-skid substance on snowy roads; and even used as cinders on some school running tracks. Perhaps as much as 20% of the total coal ash generated in the US is dumped into mineshafts as fill. In all, an estimated 40% of the coal ash generated in the US is utilized in engineering, manufacturing, agricultural and other applications.[47]

Risks of toxic exposure are least where reused coal ash is "encapsulated" or bound in a solid matrix. Examples of encapsulated uses of coal ash include as aggregate in concrete, as a component in concrete or bricks, and in the manufacture of wallboard. These applications reduce the potential threats to health, as they are more stable and less likely to leach. However, these uses may still pose a hazard to construction and demolition workers who must cut, drill or perform other dust-generating activities. Of greater concern is the reuse of coal ash in unencapsulated form, that is, as a loose particulate or sludge form that is not bound to a solid material. Common forms of unencapsulated use include using coal ash to fill mines, as fill in land contouring and construction projects, and in agriculture as a soil amendment. Severe contamination has been documented to occur as a result. In one instance, an estimated one million tons of fly ash was buried in a leaking landfill and used as construction fill in Town of Pines, Indiana. The ash contaminated drinking water wells throughout the town with toxic chemicals, including arsenic, cadmium, boron and molybdenum. Hundreds of residents were put on municipal water and the Town of Pines was declared a Superfund site, land in the US that has been contaminated by hazardous waste and is identified by the EPA as a candidate for clean-up due to the risk it poses to human health and/or the environment.[50] In another case, 1.5 million cubic yards of fly ash were used to contour a golf course in Chesapeake, Virginia; this led to groundwater contamination with arsenic, boron, chromium, copper, lead and vanadium, indicating a threat to nearby residential drinking water wells.[51]

8   Human Exposure to Coal Ash: Pathways

The standard methods of coal ash disposal are frequently insufficient to protect humans from exposure to the ash's dangerous toxicants. Exposure, direct or *via* the substances we eat, drink and breathe, has been documented to take place from all sorts of disposal sites, due to the ash or its constituent elements escaping from the sites by a variety of pathways. One of the more common is through the surface waters that flow nearby.

8.1   Surface Water Pathway

Coal ash has been documented to spill or leak into streams, rivers, ponds, lakes, wetlands and other surface waters, harming the life forms that live in and eat from those waters. The US nonprofit environmental law firm Earthjustice had documented 208 known cases of coal ash contamination and spills into bodies of water as of February 2014; however, they noted, "These cases of documented water contamination are likely to be only a small percentage of the coal ash-contaminated sites in the US. Most coal ash landfills and ponds do not conduct monitoring, so the majority of water contamination goes undetected."[45]

Catastrophic spills can take place when surface impoundment retaining walls give way, pouring enormous quantities of coal ash slurry directly into surface waters. This was the case at the coal-fired power plant in Kingston, Tennessee. Three days before Christmas 2008, an earthen wall holding back a huge coal ash pond failed. The 40-acre pond spilled more than one billion gallons of coal ash slurry into the adjacent Emory River. It covered some 300 acres with thick, toxic sludge, destroying three homes, damaging close to two dozen others, and contaminating the Emory and Clinch Rivers.[52] Fortunately, no lives were lost, perhaps because the break took place in the middle of a winter night. When the US EPA tested water samples after the spill, they found arsenic at 149 times the allowable standard for drinking water, as well as elevated levels of lead, thallium, barium, cadmium, chromium, mercury and nickel.[52]

The Kingston spill is a well-known example of failure of a coal ash pond; it is not the only case. When a dam failed at the Martin's Creek Power Plant in eastern Pennsylvania in August 2005, more than 100 million gallons of coal ash-contaminated water flowed into the Delaware River. Arsenic levels in the river jumped to levels that exceeded water quality standards and a public water supply was temporarily closed downstream. The response action cost $37 million.[46] Ash pond accidents may also occur after the electrical plant that generated them is closed. Such an accident took place in North Carolina, in the eastern United States, when in February 2014, an estimated 39 000 tons of coal ash spilled from Duke Energy's Dan River Steam Station into the Dan River near the town of Eden.[53] The steam station, or electrical power plant, had been closed for years, but its coal ash ponds had remained in place. The spill occurred, according to the well-known CBS news program "60 Minutes," when a drainage pipe, which ran under a coal ash pond and dam, collapsed. As reported by CBS News, the pipe "suck[ed] out six decades of waste and spew[ed] gunk directly into the river."[54]

The "Eden Ash Spill Site," named apparently with no intention of irony, extended some 70 miles down the Dan River from the power plant. The river itself was used for subsistence fishing, local crop irrigation and watering livestock. The spill reached the Kerr Reservoir, a recreational water body used not only for canoeing and kayaking, fishing, camping, swimming, picnicking, hiking and hunting, but also as a source of drinking water to residents in both North Carolina and Virginia. The EPA tested water in the reservoir and stated subsequently that "there have been no exceedances in human health screening in the surface water samples collected from Kerr Reservoir for contaminants of concern from the coal ash." The Agency went on to state that the drinking water samples they collected "have shown no impacts to the local drinking water" and "According to data collected, we do not believe that human health has been impacted by this coal ash spill." However, they acknowledged that they were unable to remove all of the coal ash that had been deposited in the Dan River.[55]

Small spills are far more common than these large-scale impoundment failures. Small spills may be less spectacular than a rupture, but they occur much more frequently and may continue to pollute for months or years. Small spills may occur due to leaks in impoundment dikes and dams, or overflows during heavy rains or floods. Coal ash ponds and landfills may also discharge ash-contaminated waters directly into surface waters. In fact, power plant discharges of ash-laden water – often containing very high levels of arsenic, selenium and boron – may be conducted deliberately, as a built-in aspect of disposal site operations, acknowledged and approved in operating permits. In one documented case of surface water contamination, at the US Department of Energy's Savannah River Project in South Carolina, a coal-fired power plant disposed of fly ash mixed with water into a series of open settling ponds.[56] A continuous flow of that water exited the settling ponds and entered a swamp that in turn discharged into a creek. Toxicants from the coal ash poisoned several types of aquatic animals inhabiting the wetlands: bullfrog tadpoles exhibited oral deformities and impaired swimming and predator avoidance abilities, and water snakes showed metabolic impacts. The US EPA acknowledged that these impacts were caused by releases from the ash settling ponds.

8.2 Leaching into Groundwater

In many cases, contaminants from ponds and landfills escape from their disposal sites *via* the invisible pathway known as leaching, the process by which toxics in coal ash dissolve in water and percolate through the earth. Leaching may continue to release toxic substances into the environment for decades, exposing people to dangerous toxicants at levels above safe drinking water standards. The amount of leaching that takes place at coal ash storage facilities varies greatly, reflecting the type of coal ash that is stored; its concentration and acidity; and the nature of the disposal site. As a result, leachate concentrations are different at different sites and for different elements. The rate at which leaching takes place may be affected by a number of factors: the size of the disposal pond, the pond's depth, and the amount of pressure created by the accumulated water and slurry; the types of soil and rock that lie underneath; the gradient or slope of the land; and how far beneath the pond or bottom of the landfill an aquifer or underground stream might lie.

What ultimately most determines the amount of leaching, however, is the robustness of the storage unit. The US EPA has found that two factors dramatically increase the risk that coal ash disposal units pose, both to human health and to ecosystems. The first is, quite simply, the use of wet surface impoundments rather than dry landfills. Some surface impoundments are little more than pits in the earth, with native soils as the bottom and sides; they may be contained by earthen retaining walls. These unlined wet disposal areas constitute a disproportionate number of the "damage cases" where coal ash toxics have been documented to have escaped from disposal facilities and to have damaged human health or the community;[57] see the discussion of damage cases in Section 9.

The second risk factor is the absence in surface impoundments of durable liners that can keep coal ash-laden water from accessing groundwater. To reduce the risks of leaking and leaching, some surface impoundments are lined with clay up to three feet deep. However, clay liners that are not re-inforced with other materials have been found to allow leaching of toxics into underlying groundwater. The greatest level of protection is afforded when surface impoundments are built with composite liners, constructed from various layers including human-made materials, such as a plastic membrane of high-density polyethylene, placed over clay or geosynthetic clay. According to the EPA study of projected leaching times, the median average years until peak well water concentrations are reached by selenium, arsenic and cobalt escaping from composite-lined units is in the thousands of years.[46] This indicates the vast superiority of composite-lined over unlined and clay-lined impoundments. Regardless, even composite liners have a finite lifespan and may be subject to accidents and unforeseen events, so even they cannot be said to provide truly permanent containment of coal ash toxicants.

Studies indicate that many coal ash toxicants are capable of leaching at concentrations high enough to seriously endanger human health. This was the conclusion, for example, of a laboratory study released by the US EPA in 2009.[58] The study found that for some coal ashes and under some circum-stances, the levels of toxic constituents leaching out of coal ash could reach levels hundreds to thousands of times greater than federal drinking water standards. Several toxic pollutants, including arsenic and selenium, leached in some circumstances at levels exceeding those which the US government defined as a hazardous waste. These included concentrations of arsenic of 18 000 parts *per* billion (ppb), fully 1800 times the federal drinking water standard and over three times the level that defines a hazardous waste, and selenium at 29 000 ppb, a level 580 times the drinking water standard and 29 times the hazardous waste threshold. The report observed that the leach test results represented a theoretical range of the potential concentrations of toxics that might occur in leachates, rather than an estimate of the amount of a toxic that would actually reach any given aquifer or drinking water well. At the same time, the report noted the multiple factors that affect leaching rates and concentrations; these include the pH of the ash itself, the acidity of the environment, and the variety of other conditions that coal ash encounters in the field when it is disposed or recycled. Thus, the EPA report cautioned, any evaluation using a single set of assumptions is insufficient to reflect real-life conditions and "will, in many cases, lead to inaccurate conclusions about expected leaching in the field."

A separate risk assessment study by the US EPA found that under certain circumstances, the risk of harm due to coal ash leachate reaches levels that increase the risk of cancer or other diseases.[43] At greatest risk are people who get their drinking water from a well and who live near an unlined surface impoundment that contains coal ash co-mingled with other coal wastes. According to this study, people in those conditions have as much as a one in fifty chance of getting cancer from drinking water contaminated by arsenic. This risk is 2000 times greater than the US EPA's goal for reducing cancer risk to one in 100 000.

The risk from leaching goes beyond the theoretical; verified damage from leaching has occurred at dozens of coal ash disposal sites throughout the US, contaminating drinking water, streams, and ponds and killing wildlife. In the town of Gambrills, Maryland, for instance, residential drinking wells were contaminated after fly ash and bottom ash from two Maryland power plants were dumped into excavated portions of two unlined quarries. Groundwater samples collected from residential drinking water wells near the disposal site indicated contamination with arsenic, beryllium, cadmium and lead. The EPA determined the site to be a proven damage case, as groundwater samples from residential drinking wells near the site included heavy metals and sulfates at or above groundwater quality standards.[53] Eventually, the power plant owner, Constellation Energy, settled with residents of Gambrills for $54 million for poisoning water supplies with dangerous pollutants.

8.3 Airborne Coal Ash

Coal ash can go airborne and is dangerous if inhaled. Dry disposal options such as landfills can generate dangerous quantities of this airborne ash. Known as "fugitive dust," it blows readily; the US EPA noted that in the absence of fugitive dust controls, "there is not only a possibility, but a strong likelihood" that PM levels will exceed the national ambient air quality standards for particulates.[59] To compound the problem, high background levels of PM already exist in some environments; adding fugitive dust from coal ash only serves to multiply human health risks. Protective practices to control dust, such as moistening dry coal ash or covering it, can minimize the dangers to health from this source, but such protective measures may not be required by law; in fact, many of the US states do not require daily covering of dry ash landfills, and most do not require coal ash ponds be capped to control dust. At some coal ash dump disposal sites, dust controls are applied only monthly or even yearly.

Windblown particulates of coal ash can also arise when coal ash for dry disposal is loaded and unloaded, transported, or when vehicles travel through ash disposal sites. Workers and nearby residents thus run the risk of being exposed to significant amounts of fugitive dust. Residents living near landfills may be exposed first during ash unloading and subsequently due to windblown emissions, thus probably being exposed to greater quantities dust for longer periods of time. Similarly, coal ash may blow or erode from sites where it is used for construction fill and in engineering projects, or where it is applied on agricultural fields as a soil amendment. Finally, fugitive dust can arise from wet surface impoundments in arid environments or during droughts; under those conditions they may experience significant drying, in which case the wind may cause dispersion of dried ash.

Airborne particles of coal ash, if breathed in, can affect the lungs and bronchii. Of particular concern are the extremely small particles known as $PM_{2.5}$ that can lodge deep within the lung or can pass through the lungs into the bloodstream. As with PM generated by combustion, airborne coal ash can trigger serious adverse health effects ranging from asthma attacks to increased mortality rates.

8.4   *Exposure* via *Contaminated Fish*

Human health may be harmed by eating fish from water sources contaminated by coal ash toxicants. Coal-fired power plants account for almost a third of the toxic pollution discharged to US rivers and streams from all industrial sources, according to the nonprofit organization Environmental Integrity Project.[60] Rivers, lakes and other aquatic environments can become contaminated, as we have seen, as coal ash follows any of its multiple pathways. Once the toxics are in the water or sediment, they can be absorbed by fish which absorb them through their gills or eat contaminated foods; algae, worms, and other fish food sources have all been shown to absorb coal ash toxicants. As is noted above, several of the toxic substances commonly occurring in coal ash are bioaccumulative, among them mercury and selenium. The US EPA, in modeling human exposure to mercury from contaminated fish, estimated that 65% of the bodies of water that receive discharges from power plants are associated with unsafe methylmercury ingestion.[61]

One well-documented case, that of Belews Lake, near Winston-Salem, North Carolina, serves to illustrate the dangers of fish contamination from coal ash. Belews Lake served as a cooling reservoir for a large coal-fired power plant. Fly ash produced by the power plant was disposed in a settling basin, which released selenium-laden water to the lake. Due to the selenium contamination, 16 of the 20 fish species originally present in the reservoir died out entirely, including all the primary sport fish. The state issued a health advisory in 1993, urging people to reduce their consumption of fish from the lake; it was maintained in effect for seven years, years after wastewater dumping was discontinued.[56]

8.5   *Exposure* via *Drinking Water*

Drinking water can contain coal ash toxics at levels that create risks of cancer and neurological harm. The US EPA has at times downplayed the likelihood of dangerous levels of toxics being found in drinking water, stating that in the United States "public drinking water supplies are already treated for pollutants that pose human health risks."[62] However, toxic threats may in fact occur in drinking water, even after purification treatment. Multiple factors contribute to that threat. They were described in a study of coal ash contaminants in US waterways, conducted jointly by staff working for several nonprofit organizations, including one of the authors of this chapter (Gottlieb).[62] First, although pollution standards exist for drinking water for a number of individual pollutants, standards are sometimes violated; when that happens, violations may result in millions of people being exposed to higher-than-authorized levels of toxic substances. In addition, legal standards may be set at levels that allow for significant risks to health. The maximum contaminant level for arsenic, for example, is not set solely at levels that would protect human health; rather, the standard is set at a less stringent level so as to spare local jurisdictions the need to engage in more costly treatment processes.[62]

In addition, legal standards simply do not exist for a number of the toxic substances associated with coal ash. This means that drinking-water utility companies are not required either to test for those substances or to remove them from the water supply. One example is manganese, which can cause damage to the developing nervous system and which US power plants discharge at a rate of over 14 million pounds *per* year.[62] Because no legal limit is set on manganese in US drinking water, consumers of that water not going to be made aware if they are ingesting significant quantities of manganese; in fact, in all likelihood, they have no idea there is reason to be aware. Finally, contaminant levels are set for individual toxicants; the possibility of concurrent exposure to several contaminants, and resulting synergistic interactions, is not even considered. Taken together, these factors illustrate that even fully treated drinking water that meets government standards for purity cannot be assured of being free of dangerous and damaging substances from coal ash.

9 Coal Ash "Damage Cases"

Risks and long-term concerns may be dismissed as theoretical; however, hundreds of cases exist where coal ash has been documented to cause actual harm. US law requires the EPA to examine cases of the disposal of coal combustion wastes in which danger to human health or the environment has been proved. Such determinations are not made lightly; for "proven damage" to be found, evidence must show one or more of the following: either a toxic substance has been found and measured in ground water at levels above the highest level of a contaminant allowed in drinking water; these toxics have been found at a distance from the waste storage unit "sufficient…to indicate that hazardous constituents have migrated to the extent that they could cause human health concerns;" a scientific study has provided documented evidence of another type of damage to human health or the environment; or an administrative ruling or court decision presents an explicit finding of specific damage to human health or the environment.[46] Where the finding of proven damage is made, the EPA can require corrective measures to be taken, such as closure of the unit; capping the unit; installation of new liners; groundwater treatment; groundwater monitoring; or combinations of these measures. However, remediation may not be possible; the contaminants that have already escaped into the surrounding environment may remain to do serious harm to nearby populations.

There were 208 known cases of documented water contamination from coal ash as of early 2014, according to Earthjustice, a nonprofit environmental law organization that has studied coal ash extensively and litigated on the issue.[45] However, as that organization goes on to note, "Most coal ash landfills and ponds do not conduct monitoring, so the majority of water contamination goes undetected." A few examples give a sense of the severity that damage cases can entail. These cases are among 67 proven or potential damage sites identified by the US EPA in 2007.[63]

Residential wells contaminated with multiple toxic metals. For 20 years, the disposal site for wastes from the Yorktown Power Station in the eastern state of Virginia stored fly ash from coal and petroleum coke in abandoned sand and gravel pits. Six years after the last load of coal ash was disposed of, area residents reported that the water in their drinking wells had turned green. Studies found their wells to be contaminated with nickel, vanadium, arsenic, beryllium, chromium, copper, molybdenum and selenium. Fifty-five homes had to be placed on public water, as their well water was too dangerous to drink. This site was listed on the nation's list of most polluted Superfund sites.

Leaking unlined coal ash pond contaminates drinking wells, ranches. At the PPL Montana Power Plant in Colstrip, Montana, in the wide-open spaces of the Far West, leaking unlined coal ash ponds contaminated drinking water wells with high levels of dissolved boron and sulfate. The community located nearest the power plant had to be supplied with safe drinking water. The plume of contamination stretched at least a mile from the power plant, affecting ranchers far from the waste ponds.

Landfill contaminates wells with lead, a potent neurotoxicant. In New York State, a leaking solid waste management facility containing fly ash, bottom ash and other material generated by the Dunkirk Steam Station contaminated drinking water wells with lead, a neurotoxicant that can harm the developing nervous system even at low levels of exposure. The facility was ordered to close, and ground water and surface water monitoring and maintenance were expected to continue for 30 years.

Unpredictable failure of storage ponds. At the United Power Coal Creek Station in the northern plains state of North Dakota, evaporation ponds and ash disposal ponds were built with protective linings. However, the linings of several impoundments developed severe leaks within a few years of construction. Ground water monitoring at the site showed arsenic and selenium in excess of health-based levels. The state eventually required that the ponds be relined with a composite liner.

10   Protecting Human Health from Coal Ash

10.1   *Vulnerable Populations*

The dangers associated with exposure to coal ash do not fall equally on all people. Children are particularly susceptible to harm from toxic exposures, due to windows of susceptibility, physiological differences and unique behaviors. Children's organ systems, particularly the nervous system, are still undergoing development and are thus more susceptible than those of adults to severe and permanent harm from toxic effects. They are most vulnerable during gestation (*in utero*) and infancy, and remain susceptible throughout childhood. Size also affects vulnerability: any given exposure is more significant for a child's small body than the same exposure would be for a much larger adult. Children also tend to spend more time out-of-doors, breath more rapidly than adults, and their lungs are proportionately larger, thus increasing their exposure to airborne toxics. Finally, young children are prone to hand-to-mouth behaviors that expose them to higher levels of ingested contaminants, as can be the case with fugitive dust from exposed dry coal ash.

Proximity is also an important factor in vulnerability to coal ash toxics. Many coal ash disposal sites are located in rural areas, where land availability, lower land prices and lower levels of citizen efficacy in opposing powerful economic interests make it practicable for utility companies to purchase the multi-acre sites necessary for ash ponds and landfills. Most coal ash disposal sites are on the power plant site, thus avoiding costly transportation of the ash, but concentrating the pollution. In the US, racial minorities and low-income communities live near coal ash disposal facilities in disproportionate numbers.[53,63] This proximity means greater risk of both airborne and water-borne exposure. In practice, this often translates into higher health risks for communities that, due to socioeconomic status, are already likely to bear heavier disease burdens and to enjoy fewer resources for addressing them: health insurance, disposable income, good nutrition, adequate transportation, and political and financial clout, to name a few. For this reason, coal ash exposure often becomes an environmental justice issue.

The toxic burden of coal ash falls on vulnerable low-income communities and communities of color in another way as well: by contaminating the bodies of water where these populations fish for subsistence. Many African American and Native communities rely on fishing to supply or supplement their basic nutritional needs. Traditionally, fishing provided a convenient, inexpensive and healthful food source; however, when fish are contaminated, these benefits are offset by increased health risks. The US EPA has documented the fact that disproportionately high concentrations of people of color and low-income people live not only near, but also immediately downstream from coal ash impoundments.[62]

10.2   Best Available Technologies

Coal ash multiplies the threats to health that arise from coal combustion, exposing people to multiple toxic substances and extending the threat both geographically and temporally. Yet affordable technologies can significantly reduce those threats. Disposal of coal ash can maximize protection by adhering to prudent, preventative options, including:

- Incorporating the best available elements of preventative hazard design in storage and disposal facilities. These include: engineered composite liner systems, leachate collection systems, long-term ground water monitoring and effective corrective action, should these systems fail.
- Phasing out the wet storage of coal ash, the disposal of coal ash in mines and unprotected landfills, and the reuse of unencapsulated ash where it is exposed to surface or ground water.
- Pursuing further independent research and assessment of coal ash recycling. Reuse of coal ash should be permitted only where research indicates that the toxic chemicals in coal ash will not migrate from the ash in quantities that pose a threat to human health or the environment during the entire lifecycle of the reuse application.
- Conducting research to determine the possible health effects from coal combustion waste on workers exposed to coal ash and sludge at disposal facilities, construction projects and manufacturing plants.

10.3   Coal Ash Regulation

In the United States, federal regulation has allowed some gains in protecting human health and the environment from coal ash contamination. In 2015, the US EPA finalized a regulation designed to control toxic discharges into waterways from coal-fired power plants, air-pollution control devices called scrubbers, and coal ash dumps. These Effluent Limitations Guidelines prohibited the discharge of ash wastewater and set new restrictions on arsenic, mercury, selenium and nitrate concentrations in the wastewater from air pollution scrubbers. The EPA expected the new standards, which were scheduled to take effect in 2018, to eliminate over 90% of these and other toxics like cadmium, chromium and lead that coal-fired generators pour into US waterways every year. However, a nonprofit environmental law organization warned that a "backlog of expired state permits and weaknesses in monitoring requirements could delay or derail implementation" of these potentially transformative regulations.[64] In any case, in 2017 the EPA under the Trump administration chose to delay implementation of the Effluent Limitations Guidelines, and at the time this book went to press their fate was far from sure.

In 2016 the EPA completed a regulation specifically concerning disposal of coal ash from power plants.[65] The regulation established technical requirements for coal ash landfills and surface impoundments, and established recordkeeping and reporting requirements as well as requirements for each facility to post specific information to a publicly accessible website. However, it addressed coal ash under the nation's law for regulating solid waste, as opposed to hazardous waste, and left enforcement primarily to the states. Specifically, the regulation sets guidelines for regulating coal ash but allows states to opt out of those guidelines. It then defines the facilities that fail to follow those optional guidelines as "open dumps" and gives citizens the right – in effect the responsibility – to file suit to enforce legal provisions against open dumps. This weakness in regard to enforcement undercuts the value of the regulation. It places the burden on citizens to file law suits to enforce the law, after the law has been violated. In effect, it forces citizens to combat the utility companies in order to gain protection from a massive waste stream.

The burden from this regulation may fall most heavily on vulnerable communities. This was noted in a report issued by the US Civil Rights Commission in 2016, stating "The EPA's Final Coal Ash Rule negatively impacts low-income and communities of color disproportionately, and places enforcement of the Rule back on the shoulders of the community. This system requires low-income and communities of color to collect complex data, fund litigation and navigate the federal court system..."[62] That burden may well exceed the capacities of many affected communities to defend themselves.

While the use of the best available engineering safeguards can significantly increase the efficacy of coal ash containment, the threats can never be perfectly and permanently contained. Furthermore, although governments can establish enforceable safeguards that protect health and the environment, they may fail to do so, or those safeguards may prove to be insufficient, or may be flouted. As a result, many communities can expect to experience increased levels of human exposure to coal ash toxics in the future. This dangerous and unnecessary threat provides yet another health-based argument for reducing reliance on coal as a means of generating electricity.

11 Coal and Climate Change

In December, 2015, nearly 200 nations gathered together in Paris at the 2015 United Nations Climate Change Conference, sometimes known as COP 21. The goal was to adopt a universal agreement to combat climate change. The resulting agreement was signed by 174 nations on Earth Day, April 22, 2016. This followed the near-universal agreement by national academies of science in nations around the world and most notably in statements published in the Fifth Assessment Report of the Intergovernmental Panel on Climate Change.[66] The assessment of the IPCC was emphatic and stark. It stated that "warming of the climate system is unequivocal... the largest contribution is from CO_2 [and] continued emissions of greenhouse gases will cause further warming." The forcefulness of their language was unusual for scientists, who typically make more nuanced statements.

11.1   Coal and Greenhouse Gases

Carbon dioxide is the most important of the greenhouse gases in the atmosphere. Its importance is related to its concentration, the fact that it has risen at an unprecedented rate, beginning at the onset of the industrial revolution; its high degree of radiative forcing (a measure of its greenhouse effect) and the fact that it has an extraordinarily long lifetime in the environment.[31] The present atmospheric concentration of 400 ppm is certain to remain at that level for the foreseeable future and is higher than at any time in the past 24 million years.[67] Burning coal is the major contributor to this rise.

According to the Emissions Data Gathering, Analysis and Retrieval system (EDGAR), a service of the European Union, the total amount of carbon dioxide emitted into the atmosphere topped 36 billion tons in 2015, the most recent year for which data are available.[68] China and the US are the two leading emitters. Generating electricity is by far the largest contributor, accounting for 73% of the emissions, according to the Department of Energy (DOE).

The Fifth IPCC Assessment refers to our time as "The era of climate options."[66] The Commission on Climate Change and Health took a similarly cautious approach when it wrote that "Tackling climate change could be the greatest public health opportunity of the 21st century".[69] As temperatures rise without effective efforts to curb greenhouse gas emissions, options diminish along with opportunities. Recent data paint a somewhat grim picture. Global surface temperature for the earth showed that the 2011–2015 interval was the hottest ever recorded.[70] Since then, record-breaking temperatures have continued.

11.2   Climate Change and Health

The IPCC presented a semi-quantitative depiction of the impacts of climate change on various domains of health along with the effects that are possible presuming that there is an aggressive public health response and adaptation. This is shown in Figure 4.

Undernutrition is perhaps the most important global effect. This is reflected by the WHO report that there were more than 250 000 excess deaths due to starvation in the Horn of Africa in the years 2011 and 2012. The reasons for this are complex (for review, see ref. 32.). Drought is the major factor, causing the crop failures that lead to starvation. In addition, the yields of most agricultural commodities are sensitive to temperature. As temperatures rise above critical thresholds, critical stages in the production of the crop may be exceeded, causing crop failure. This is often during the process of fertilization. Rising atmospheric carbon dioxide levels may also have an impact. The nutritional value of the commodity may fall when carbon dioxide levels rise. In addition, rising carbon dioxide levels may favor the growth of weeds over the growth of the commodity. Climate change has created a perfect storm of converging problems that act synergistically, often

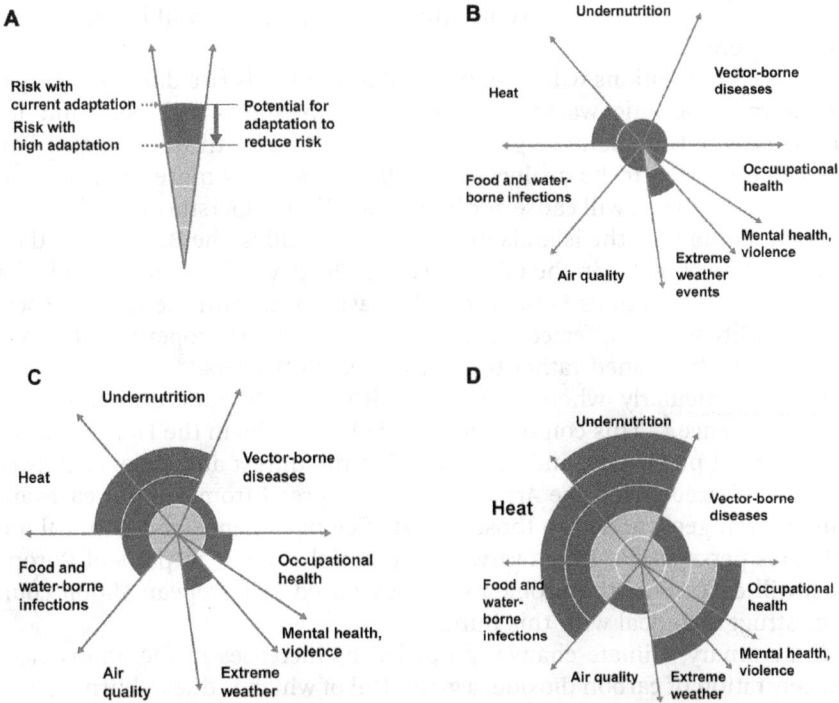

Figure 4 Targeting health. Panel A is the key to understanding the remainder of the figure. The magnitude of the risk for each factor is shown by the width of the slice of the pie. The darkened portion of each slice depicts the potential for risk reduction in a hypothetical, highly adapted condition. Panel B shows the relative importance of the burden of poor health at present in a qualitative way. Panel C depicts the risks of and potential benefits to be gained from adaptation in the relatively near term, 2030 to 2040. Panel D portrays the relative risks and adaptation potentials toward the end of the century, 2080 to 2100, with a temperature rise of 4 °C relative to the preindustrial era.
(Data taken from *Working Group II Contribution* to *the Fifth Assessment Report of the Intergovernmental Panel on Climate Change.*[76])

along with the burden of diseases such as malaria, to imperil health on much of the planet.

Heat and heat-related illnesses, such as heat exhaustion and the more severe heat stroke, will become increasingly common. A European heat wave in 2003 caused about 70 000 deaths, and more than 4100 deaths occurred in Pakistan and India in 2015.[71,73] Rising temperatures are certain to make this bad problem worse. Fortunately, at least for the developed world, public health responses centered on moving susceptible individuals to cooling shelters and providing air conditioning and other forms of support to isolated individuals and sensitive populations can diminish the impact of heat waves.[31] Most people who live in Third World countries will be left to fend for themselves.

Some island nations will be obliterated as sea levels rise due expansion of the warmer oceanic water combined with melting of glaciers and ice, particularly in Greenland. High sea levels, combined with storm surges from what is predicted to be an increase in the number of more intense hurricanes and cyclones, will cause additional flooding. Superstorm Sandy, which devastated many of the islands in the Greater Antilles, the Bahamas and the Middle Atlantic states in the US in the fall of 2012, was just a preview of what could come.[72] Again, as is too often the case, those with the highest social vulnerability will be affected the most severely – their property is the most likely to be abandoned rather than protected by barriers.[73]

Heat, particularly when combined with drought and undernutrition, breeds violence.[31] This combination has led to deaths in the Horn of Africa, as described previously, and contributed to the unrest and violence in Syria and that characterized the Arab Spring as it spread from Tunisia eastward. Climate refugees, added to those already fleeing violence and political and religious persecution, have overwhelmed social systems in parts of Europe. This will exacerbate the problems already posed as European Union countries struggle to deal with this burden.

In summary, climate change, propelled by increases in the atmospheric concentration of carbon dioxide, a great deal of which is due to burning coal, is a major threat to all of us. The political leadership needed to deal with this problem effectively is fragmentary and totally lacking in some circles of power. More reliance on, and belief in, peer-reviewed science, with evidence-based decision-making backed by firm support from all stakeholders, is needed to mitigate and adapt to the changing climate if we are to avoid the severe threats to health and survival that await an unprepared world.

12   Coal, Human Health and the Precautionary Principle

As this chapter documents, the use of coal as a power source results in severely damaging effects on human health and the environment. For this reason, coal must be addressed not just as a source of energy but as a public health issue as well. Reliance on coal is also an ethical issue. When a human process disperses arsenic, lead, mercury, selenium and a dozen other toxicants into the environment, it robs us and our children of our potential for full development. It also harms the much broader biotic community. When that process hastens the heating of the atmosphere and the oceans, it puts the entire planet at risk.

The corporations that burn coal remain in many cases free of responsibility for the consequences that their activities unleash on human and planetary health. From a medical perspective, this situation calls for application of the "precautionary principle." This principle states "Where an activity raises threats of harm to the environment or human health, precautionary measures should be taken even if some cause and effect relationships are not fully established scientifically."[74] Further application of the precautionary principle proposes that where an action risks causing harm to the public or to the environment, the burden of proof that it is not harmful falls on those who would take the action.[75] In contrast to a classical risk assessment approach, which asks "How much harm can we tolerate?", the precautionary principle asks "What actions can we take to prevent harm?" Viewed through the lens of the precautionary principle, it is clear that those who produce, utilize and dispose of coal and coal waste products should assume the responsibility of protecting people and the environment from the resulting pollution and harm.

References

1. Mine Safety and Health Administration, *Performance Coal Company Upper Big Branch Mine-South, Massey Energy Company*, 2014, Mine ID: 46-08436 and news release.
2. US Department of Labor, Bureau of Labor Statistics, National Census of Occupational Injuries 2015, https://www.bls.gov/news.release/pdf/cfoi.pdf. Accessed 30 Mar 2017.
3. US National Library of Medicine, *Medline Plus*, https://medlineplus.gov/ency/article/000130.htm. Accessed 29 Jan 2017.
4. A. Makellar, An Investigation into the Nature of Black Phthisis/or Ulceration Induced by Carbonaceous Accumulation in the Lungs of Coal Miners (Kindle Locations 43-46).
5. Centers for Disease Control, *Pneumoconiosis Prevalence Among Working Coal Miners Examined in Federal Chest Radiograph Surveillance Programs - United States, 1996-2002, Morbidity & Mortality Weekly Report*, 2003, **52**(15), 336.
6. A. S. Laney and M. D. Attfield, Coal workers' pneumoconiosis and progressive massive fibrosis are increasingly more prevalent among workers in small underground coal mines in the United States, *Occup. Environ. Med.*, 2010, **67**, 428.
7. Lowering Miners' Exposure to Respirable Coal Mine Dust, Including Continuous Personal Dust Monitors. 79 FR 24813 (2014). https://www.federalregister.gov/documents/2014/05/01/2014-09084/loweringminers-exposure-to-respirable-coal-mine-dust-including-continuouspersonal-dust-monitors. Accessed 1 Jan 2017.
8. D. J. Blackley, C. N. Halldin and A. S. Laney, Resurgence of a debilitating and entirely preventable respiratory disease among working coal miners, *Am. J. Respir. Dis. Crit. Care Med.*, 2014, **190**, 708.
9. E. S. Bernhardt, B. D. Lutz, R. S. King, J. P. Fay, C. E. Carter, A. M. Helton, D. Campagna and J. Amos, How many mountains can we mine? Assessing the regional degradation of central Appalachian rivers by surface coal mining, *Environ. Sci. Technol.*, 2012, **46**, 8115.
10. M. Hendryx and M. M. Ahern, Relations between health indicators and residential proximity to coal mining in West Virginia, *Am. J. Public Health*, 2008, **98**(4), 669.

11. M. Hendryx, Mortality rates in Appalachian coal mining counties: 24 years behind the nation, *Environ. Justice*, 2008, **1**, 5.
12. J. Leonard, *Coal Preparation*, Society for Mining, Metallurgy, and Exploration, Littleton, CO, 5th edn, 1991.
13. West Virginia Division of Culture and History, *Buffalo Creek*, 2010, West Virginia Archives and History.
14. Martin County Coal Corp., Coal Slurry Release Work Plan, 2001, Ecology and Environment, Lancaster, NY.
15. D. Lovan, Inez Coal Slurry Spill: Toxic Sludge From Massey Facility Still Pollutes Kentucky Town A Decade After Disaster, *Huffington Post*, 10 October 2010.
16. A. H. Lockwood, *The Silent Epidemic: Coal and the Hidden Threat to Health*, MIT Press, 2012.
17. J. Eilperin and S. Mufson, Many coal sludge impoundments have weak walls, federal study says, *Washington Post*, 24 April 2013.
18. Committee on Coal Waste Impoundments, Committee on Earth Resources, Board on Earth Sciences and Resources Staff and National Research Council Staff, *Coal Waste Impoundments: Risks, Responses, and Alternatives*, National Academies Press, 2002.
19. Association of American Railroads, *Railroads and Coal*, July 2016, https://www.aar.org/BackgroundPapers/Railroads%20and%20Coal.pdf. Accessed 5 Mar 2017.
20. United States Environmental Protection Agency, *EPA Finalizes More Stringent Emissions Standards for Locomotives and Marine Compression-Ignition Engines*, March 2008. https://nepis.epa.gov/Exe/ZyPDF.cgi/P100094D.PDF?Dockey=P100094D.PDF. Accessed 16 Feb 2017.
21. United States Environmental Protection Agency, Regulations for Emissions from Locomotives, https://www.epa.gov/regulations-emissions-vehicles-and-engines/regulations-emissions-locomotives. Accessed 16 Feb 2017.
22. Donora Smog Museum, http://www.donorasmog.com. Accessed 28 Jan 2017.
23. N. Bruce, R. Perez-Padilla and R. Albalak, Indoor air pollution in developing countries: a major environmental and public health challenge, *Bull. W. H. O.*, 2000, **78**, 1078.
24. United States Environmental Protection Agency, Study of Hazardous Air Pollutant Emissions from Electric Utility Steam Generating Units - Final Report to Congress, *EPA publication 453/R-98-004a*, 1998.
25. www.enviroflash.info. Accessed 30 Mar 2017.
26. G. A. Wellenius, M. R. Burger, B. A. Coull, J. Schwartz, H. H. Kuh, P. Koutrakis, G. Schlaug, D. R. Gold and M. A. Mittleman, Ambient air pollution and the risk of acute ischemic stroke, *Arch. Intern. Med.*, 2012, **172**, 229.
27. J. Schwartz, D. W. Dockery and L. M. Neas, Is daily mortality associated specifically with fine particles?, *J. Air Waste Manage. Assoc.*, 1996, **46**, 927.

28. F. Laden, L. M. Neas, D. W. Dockery and J. Schwartz, Association of fine particulate matter from different sources with daily mortality in six US cities, *Environ. Health Perspect.*, 2000, **108**, 941.
29. G. D. Thurston, R. T. Burnett, M. C. Turner, Y. Shi, D. Krewski, R. Lall, K. Ito, M. Jerrett, S. M. Gapstur, W. R. Diver and C. A. Pope III, Ischemic heart disease mortality and long-term exposure to source-related components of US fine particle air pollution, *Environ. Health Perspect.*, 2016, **124**, 785.
30. M. L. Block and L. Calderón-Garcidueñas, Air pollution: mechanisms of neuroinflammation and CNS disease, *Trends Neurosci.*, 2009, **32**, 50.
31. H. J. Heusinkveld, T. Wahle, A. Campbell, R. H. S. Westerink, L. Tran, H. Johnston, V. Stone, F. R. Cassee and R. P. F. Schins, Neurodegenerative and Neurological Disorders by Small Inhaled Particles, *NeuroToxicology*, 2016, **56**, 94.
32. A. H. Lockwood, *Heat Advisory: Protecting Health on a Warming Planet*, MIT Press, Cambridge, MA, 2016.
33. M. L. Bell, R. Goldberg, C. Hogrefe, P. L. Kinney, K. Knowlton, B. Lynn, J. Rosenthal, C. Rosenzweig and J. A. Patz, Climate Change, Ambient Ozone and Health in 50 US Cities, *Clim. Change*, 2007, **82**, 61.
34. S. Wu, L. J. Mickley, D. J. Jacob, D. Rind and D. G. Streets, Effects of 2000–2050 changes in climate and emissions on global tropospheric ozone and the policy-relevant background surface ozone in the United States, *J. Geophys. Res.: Atmos.*, 2008, **113**, D18312.
35. C.-W. Yap, R. Hoyle and A. Ismar, China to Ban Coal with High Ash, Sulfur, *Wall Street Journal*, New York, NY, 14 September 2014.
36. UNEP, *Global Mercury Assessment 2013: Sources, Emissions, Releases and Environmental Transport*, UNEP Chemicals Branch, Geneva, Switzerland, 2013.
37. C. T. Driscoll, Y. J. Han, C. Y. Chen, D. C. Evers, K. F. Lambert, T. M. Holsen, N. C. Kamman and R. K. Munson, Mercury Contamination in Forest and Freshwater Ecosystems in the Northeastern United States, *BioScience*, 2007, **57**, 17.
38. US EPA, Fish and Shellfish Advisories and Safe Eating Guidelines, https://www.epa.gov/choose-fish-and-shellfish-wisely/fish-and-shellfish-advisories-and-safe-eating-guidelines. Accessed 25 Jan 2017.
39. Physicians for Social Responsibility, Healthy Fish, Healthy Families, http://www.psr.org/resources/healthy-fish-healthy-families.html. Accessed 25 Jan 2017.
40. World Health Organization, *Media Centre: Mercury and Health*, http://www.who.int/mediacentre/factsheets/fs361/en/ updated January 2016. Accessed 25 Jan 2017.
41. National Research Council, *Hidden Costs of Energy: Unpriced Consequences of Energy Production and Use*, National Academy of Sciences, Washington, D.C., 2009.
42. P. R. Epstein, J. J. Buonocore, K. Eckerle, M. Hendryx, B. M. Stout, R. Heinberg, R. W. Clapp, B. May, N. L. Reinhart, M. M. Ahern,

S. K. Doshi and L. Glustrom, Full cost accounting for the life cycle of coal, *Ann. N. Y. Acad. Sci.*, 2011, **1219**, 73.

43. U.S. Environmental Protection Agency Office of Air and Radiation, *The Benefits and Costs of the Clean Air Act: 1990 to 2020*, U.S. EPA, Washington D.C., 2010.

44. U.S. Environmental Protection Agency, Office of Solid Waste and Emergency Response, Office of Resource Conservation and Recovery, *Human and Ecological Risk Assessment of Coal Combustion Wastes Draft EPA document*, April 2010, 2.3–2.5. http://earthjustice.org/sites/default/files/library/reports/epa-coal-combustion-waste-risk-assessment.pdf. Accessed 21 Feb 2017.

45. U.S. Environmental Protection Agency (1999), *Report to Congress, Wastes From the Combustion of Fossil Fuels. Volume 2—Methods, findings, and recommendations*, Office of Solid Waste and Emergency Response, Washington, DC. EPA 530-R-99-010. March 1999.

46. Earthjustice, *Coal Ash Contaminated Sites & Hazard Dams*, http://earthjustice.org/features/map-coal-ash. Accessed 21 Feb 2017.

47. U.S. Environmental Protection Agency. *EPA-HQRCRA-2009-0640-0078, Proposed Rules*, Federal Register, Vol. 75, No. 118 / Monday, June 21, 2010, 35145. http://www.epa.ohio.gov/Portals/32/pdf/EPA-HQ-RCRA-2009-0640-0352%5B1%5D.pdf. Accessed 21 Feb 2017.

48. B. Breen, *Testimony delivered to Committee on Transportation and Infrastructure, Subcommittee on Water Resources and the Environment, U.S. House of Representatives*, April 30, 2009, https://archive.epa.gov/epawaste/nonhaz/industrial/special/fossil/web/pdf/coalashtest409.pdf. Accessed 21 Feb 2017.

49. U.S. Environmental Protection Agency, Fact Sheet: Coal Combustion Residues (CCR)—Surface Impoundments with High Hazard Potential Ratings, EPA530-F-09-006. June 2009 (updated August 2009). https://nepis.epa.gov/Exe/tiff2png.cgi/P10048EX.PNG?-r+75+-g+7+D%3A%5CZYFILES%5CINDEX%20DATA%5C06THRU10%5CTIFF%5C00000453%5CP10048EX.TIF. Accessed 21 Feb 2017.

50. U.S. Environmental Protection Agency, EPA Superfund Program: Town of Pines Groundwater Plume, Town of Pines, IN, https://cumulis.epa.gov/supercpad/cursites/csitinfo.cfm?id=0508071. Accessed 21 Feb 2017.

51. Virginia Conservation Network, *Virginia's Toxic Coal Ash Problem*, pp. 7–8. Available at http://www.cleanwateraction.org/files/publications/chesapeake/VA%20Coal%20Ash%20Report.5.13.2015.pdf. Accessed 27 July 2017.

52. Smith, Testimony of Stephan A. Smith, DVM, Executive Director, Southern Alliance for Clean Energy. Submitted to the U.S. Senate Committee on Environment and Public Works, January 8, 2009. https://www.epw.senate.gov/public/_cache/files/e918d2f7-9e8b-411e-b244-9a3a7c3359d9/testimonyofstephensmith182009.pdf. Accessed 22 Feb 2017.

53. U.S. Environmental Protection Agency, *Enforcement Case Summary: Duke Energy Agrees to $3 Million Cleanup for Coal Ash Release in the Dan River*,

https://www.epa.gov/enforcement/case-summary-duke-energy-agrees-3-million-cleanup-coal-ash-release-dan-river. Accessed 15 Feb 2017.

54. CBS News, *The Spill at Dan River*, http://www.cbsnews.com/news/duke-energy-on-coal-ash-waste-at-dan-river-2/. Accessed 22 Feb 2017.

55. U.S. Environmental Protection Agency, *Frequently Asked Questions (FAQs) about the Duke Energy Coal Ash Spill in N. C. Eden*, https://www.epa.gov/dukeenergy-coalash/frequently-asked-questions-faqs-about-duke-energy-coal-ash-spill-eden-nc. Accessed 22 Feb 2017.

56. Earthjustice, *South Carolina Coal Ash Disposal in Ponds and Landfills*, http://earthjustice.org/sites/default/files/sc-coal-ash-factsheet0811.pdf. Accessed 22 Feb 2017.

57. U.S. Environmental Protection Agency, *Office of Solid Waste. Coal Combustion Waste, Damage Case Assessments*, July 9, 2007. http://graphics8.nytimes.com/packages/pdf/national/07sludge_EPA.pdf. Accessed 22 Feb 2017.

58. U.S. Environmental Protection Agency, Office of Research and Development, D. Kosson, F. Sanchez, P. Kariher, L. H. Turner, R. Delapp and P. Seignette, *Characterization of Coal Combustion Residues from Electric Utilities—Leaching and Characterization Data*, EPA-600/R-09/151, 2009, http://www.vanderbilt.edu/leaching/assets/publications/EPA-600_R_09_151%20(Report%203).pdf. Accessed 22 Feb 2017.

59. U.S. Environmental Protection Agency, *Inhalation of Fugitive Dust: A Screening Assessment of the Risks Posed by Coal Combustion Waste Landfills*, May 2010. https://www.efis.psc.mo.gov/mpsc/commoncomponents/viewdocument.asp?DocId=935784779. Accessed 22 Feb 2017.

60. Environmental Integrity Project, *Toxic Wastewater from Coal Plants.* August 2016, http://www.environmentalintegrity.org/wp-content/uploads/Toxic-Wastewater-from-Coal-Plants-2016.08.11-1.pdf. Accessed 22 Feb 2017.

61. U.S. Environmental Protection Agency. *Effluent Limitations Guidelines and Standards for the Steam Electric Power Generating Point Source Category; Proposed Rule, 78 Fed. Reg. 34432, 34435 (June 7, 2013)*, cited in: B. Gottlieb, A. Russ, C. Roberts, L. Evans, T. Cmar and J. Peters, *Selling Our Health Down the River: Why EPA Needs to Finalize the Strongest Rule to Stop Water Pollution from Power Plants*, June 2015, http://www.psr.org/assets/pdfs/selling-our-health-down-the-river.pdf. Accessed 23 Feb 2017.

62. U.S. Environmental Protection Agency, *National Primary Drinking Water Regulations – Arsenic and Clarifications to Compliance and New Source Contaminants Monitoring*, 66 Fed. Reg. 6976, 2001. https://www.federalregister.gov/documents/2001/01/22/01-1668/national-primary-drinking-water-regulations-arsenic-and-clarifications-to-compliance-and-new-source. Accessed 23 Feb 2017.

63. U.S. Commission on Civil Rights, *Environmental Justice: Examining The Environmental Protection Agency's Compliance and Enforcement of Title VI and Executive Order 12.898*, September 2016, http://www.usccr.gov/pubs/Statutory_Enforcement_Report2016.pdf. Accessed 22 Feb 2017.

64. Environmental Integrity Project, *Toxic Wastewater from Coal Plants, August 2016*, http://www.environmentalintegrity.org/wp-content/uploads/Toxic-Wastewater-from-Coal-Plants-2016.08.11-1.pdf. Accessed 22 Feb 2017.

65. U.S. Environmental Protection Agency, *Final Rule: Disposal of Coal Combustion Residuals from Electric Utilities*, https://www.epa.gov/coalash/coal-ash-rule#summary. Accessed 22 Feb 2017.

66. IPCC, *Fifth Assessment Report of the Intergovernmental Panel on Climate Change*, Geneva, Switzerland, 2014.

67. Electronic Data Gathering, Analysis, and Retrieval system (EDGAR), http://edgar.jrc.ec.europa.eu/overview.php?v=CO2ts1990-2015. Accessed 27 Jan 2017.

68. Emissions Database for Global Atmospheric Research. Available at https://www.eea.europa.eu/themes/air/links/data-sources/emission-database-for-global-atmospheric.

69. N. Watts, W. N. Adger, P. Agnolucci, J. Blastock, P. Bayass, W. Cai, S. Chaytor, T. Colbourn, M. Collins, A. Cooper and P. M. Cox, Health and climate change: policy responses to protect public health, *Lancet*, 2015, **385**, 1.

70. World Meteorological Organization, *The Global Climate in 2011–2015*, WMO-No. 1179, World Meteorological Organization, Geneva, Switzerland, 2016.

71. J. M. Robine, S. L. Cheung, S. le Roy, H. Van Oyen, C. Griffiths, J. P. Michel and F. R. Hermann, Death Toll Exceeded 70,000 in Europe During the Summer of 2003, *C. R. Biol.*, 2008, **331**, 171.

72. *Hurricane Sandy*, https://en.wikipedia.org/wiki/Hurricane_Sandy. Accessed 7 Mar 2017.

73. J. Martinich, J. Neumann, J. Ludwig and L. Jantarasami, Risks of Sea Level Rise to Disadvantaged Communities in the United States, *Mitigation Adaptation Strategies for Global Change*, 2013, **18**, 169.

74. *Wingspread Statement on the Precautionary Principle*, http://www.gdrc.org/u-gov/precaution-3.html. Accessed 22 Feb 2017.

75. S. G. Gilbert, *Public Health and the Precautionary Principle*, Northwest Public Health, University of Washington School of Public Health & Community Medicine, Spring/Summer 2005. https://www.tpchd.org/files/library/427d8e6368fb4b6c.pdf. Accessed 22 Feb 2017.

76. D. Campbell-Lendrum, D. Chadee, Y. Honda and multiple others. "Human Health: Impacts, Adaptation, and Co-Benefits," in *Climate Change 2014: Impacts, Adaptation, and Vulnerability; Part A: Global and Sectoral Aspects; Working Group II Contribution to the Fifth Assessment Report of the Intergovernmental Panel on Climate Change*, 709–754, Cambridge University Press, New York, 2014.

The State of Coal Regulation Around the World: Insights from the United States, China, Germany and India

KEN KIMMELL* AND RACHEL CLEETUS

ABSTRACT

Coal made the Industrial Revolution possible and, at the start of the 21st century, it is still a dominant source of energy globally. However, coal combustion leads to significant air and water pollution, as well as carbon dioxide emissions, which are the leading driver of climate change. The future of coal will be determined by how quickly and seriously countries work to address public health concerns and make the transition to low-carbon development. The experiences of major economies such as the United States, China, Germany and India provide useful insights into the challenges and opportunities presented by a transition away from coal to cleaner forms of energy. This chapter summarizes the current policy and regulatory context in these countries, with a special emphasis on the US.

1 Introduction

In December 2015, over 190 nations came together to sign the Paris Agreement, aimed at making deep cuts in global heat-trapping emissions to limit some of the worst impacts of climate change. Countries committed to the aim of *"holding the increase in the global average temperature to well below*

---

*Corresponding author.

Issues in Environmental Science and Technology No. 45
Coal in the 21st Century: Energy Needs, Chemicals and Environmental Controls
Edited by R.E. Hester and R.M. Harrison
© The Royal Society of Chemistry 2018
Published by the Royal Society of Chemistry, www.rsc.org

2 °C above pre-industrial levels and pursuing efforts to limit the temperature increase to 1.5 °C above pre-industrial levels." To achieve this temperature goal, the agreement lays out a long-term goal to "*achieve a balance between anthropogenic emissions by sources and removals by sinks of greenhouse gases in the second half of this century*".[1]

The Paris Agreement sets the fundamental context within which global energy choices will be made over the next few decades. Countries, businesses and consumers understand that the market is now oriented toward a low-carbon energy pathway, which has profound implications for coal's status as the dominant source of power. Coal has the highest carbon content of any fossil fuel. Combusting coal to generate power and make steel and cement is one of the biggest sources of global carbon dioxide (CO_2) emissions.[2] Coal is responsible for approximately 45% of global energy-related CO_2 emissions and is also a significant source of other harmful pollutants such as sulfur dioxide, nitrogen oxides, particulate matter and mercury.[3–5]

Coal is cheap and plentiful today. Weaning the world off it will be challenging, and it raises complex technological and socioeconomic questions. The large public health benefits of curtailing coal use, alone, make this a worthwhile endeavor. The added climate benefits seal the deal. However, this clean energy transition will require a robust set of policies and regulations. This chapter begins with an overview of market trends and regulatory policy in a select group of coal-dependent countries and goes on to address coal regulation in the United States in greater detail.

2   Global Coal Market Trends

Global coal production is currently in a period of decline, although there are questions about the long-term sustainability of this trend. In 2015 the International Energy Agency (IEA) recorded a decline of 221 Mt, the single biggest year-on-year decrease in production since it began keeping records in 1971.[6] These declines are underpinned by similar trends in major coal producing countries like China, the United States and Germany (see Table 1).

As the biggest producer and consumer of coal, changes in China's coal demand are a big driver of global trends. Motivated by public health and climate concerns, the Government of China has made public commitments to cut back on the country's coal use and has followed through with cancelations of new coal-fired power plants and the shuttering of mines and existing power plants. As a result, coal production and use has been flat or falling over the past few years. Official estimates show coal consumption down 4.7% in 2016 compared to the previous year.[7] The next biggest coal producer, the US, has seen coal production, consumptions and employment all fall by more than 10% in 2015.[8] These trends have been driven in large part by market factors, including cheap natural gas, falling renewable energy costs and mechanization of mining.

Table 1 Major Coal Producers (Mt).[a]

	2013	2014	2015
PR of China	3748.5	3640.2	3527.2
United States	903.7	918.2	812.8
India	610.0	657.4	691.3
Australia	458.4	488.8	508.7
Indonesia	489.7	484.7	469.3
Russian Federation	326.0	332.9	349.3
South Africa	256.3	260.5	252.1
Germany	191.0	186.5	184.7
Poland	142.9	137.1	135.8
Kazakhstan	119.6	114.0	107.2
Other	*728.5*	*709.2*	*670.5*
WORLD	**7974.6**	**7929.7**	**7708.7**

[a]Production includes recovered slurries and production from other sources. (Data for Australia and India are provided on a fiscal basis). (Source: International Energy Agency).

Even as coal use may be on the decline in the US and China, projections show that without policy interventions it will likely increase in many parts of Asia to meet the needs of growing economies and populations. In just fifteen years, from 2000 to 2015, Asia's share of coal demand has gone up from less than half to three-quarters of global demand, while Europe and North America's has fallen from about half to less than a quarter.[6]

Whether the world successfully transitions away from coal ultimately will be determined by policies in both developed and newly emerging economies. There are still many unanswered questions, including the following ones. Will richer nations be able to model a rapid transition to low-carbon energy while sustaining economic prosperity? Will the costs of renewable energy fall enough to make it an attractive substitute for coal in both developed and emerging economies? Will countries revoke fossil fuel subsidies that continue to provide unfair advantages to polluting forms of energy? Will richer nations help advance new and existing low-carbon technologies abroad, in keeping with their responsibilities under the global climate framework?

3   Will Climate Commitments Drive a Shift Away from Coal?

Leading up to and since the Paris Agreement was reached, there has been a sea change in climate and clean energy policies around the world. Parties to the Paris Agreement have made commitments *via* their Nationally Determined Contributions (NDCs) that outline the major actions they will take in the near-to-medium term to cut their greenhouse gas emissions. For the first time in the context of international climate negotiations both developed and developing countries offered NDCs. Commitments from major carbon-emitting nations vary in specifics, but implicitly or explicitly they signal a shift away from coal.

The US has committed to cutting its emissions 26% to 28% below 2005 levels by 2025.[9] The most recent draft US Greenhouse Gas Inventory shows that at the end of 2015 emissions were about 11% below 2005 levels, meaning that the country is approximately 40% of the way to its Paris commitment.[10] That progress has largely been driven by the power sector shift from coal to natural gas and, to a lesser extent, renewable energy. Despite attempts by the Trump administration to roll back Obama-era climate policies such as the Clean Power Plan, and its decision to withdraw from the Paris Agreement, projections show that market trends, state policies and federal renewable energy tax credits will continue to drive this shift. Nevertheless, a coal-to-gas switch is by no means sufficient to deliver on climate goals and more assertive economy-wide policies will be needed.

China's NDC aims to, by 2030 or before, achieve a peaking of the country's carbon dioxide emissions, lower the carbon intensity of its GDP by 60% to 65% from 2005 levels, and increase the share of non-fossil fuels in primary energy consumption to approximately 20%.[11] Since 2015, the country has announced a number of related domestic policy actions, and the data show impressive progress toward potentially meeting these goals well ahead of 2030. Based on China's National Energy Administration forecasts for energy consumption, the country's CO_2 emissions are projected to fall by 1% in 2017, marking the fourth year in a row that its emissions have fallen or re-mained stable.[12] Consistent with this, the IEA estimates that China's coal consumption probably peaked in 2013.[13] The country recently announced the cancellation of over 100 new coal plants, following a similar an-nouncement in 2016.[14] The country aims to invest over $360 billion in re-newable energy through 2020, and create over 13 million jobs in the sector.[15]

Germany's *Energiewende* (Energy Transition), which has its origins in a 1970s grassroots movement opposing nuclear energy, has matured under Chancellor Angela Merkel into a full-fledged plan to promote renewable energy and transition the country to a low-carbon energy pathway. A feed-in-tariff program is credited with increasing the share of renewable power from just 5% in 1999 to approximately 33% in 2016.[16] In May 2016, the country briefly produced nearly 90% of its power from renewable energy resources. However, Germany is facing some significant challenges as it tries to further ramp up renewable energy even as it tries to phase out nuclear power by 2022, a commitment made in the wake of the Fukushima disaster.[17] Recent budget problems have forced the government to abandon the feed-in-tariff program, although the 2016 Renewable Energy Sources Act aims to raise the share of renewable energy to 40–45% in 2025 and 55–60% in 2035.[18]

Coal still helps meet about 40% of Germany's energy needs.[17] Para-doxically, the country's emissions rose in 2015 as its reliance on coal, in-cluding "dirty" lignite coal, increased. The country's Climate Action Plan 2050, released in December 2016, attempts to outline a path to a 80–95% reduction in greenhouse gas emissions compared to 1990 levels by 2050.[19,20] The plan's failure to set a deadline for the phase-out of coal is seen as a weakness by many environmental groups, and resulted from very contentious political discussions about the economic implications of such a phase-out.

A number of European utilities, including from Germany, recently announced that they will not build any more new coal plants in the European Union after 2020. The press release from Eurelectric, the union of the European electricity industry, specifically cited a commitment to help meet the Paris Agreement as the reason for this decision.[21]

4   National Policies and Policy Drivers Related to Coal

Concerns about energy security and affordability, public health and climate change are all important considerations for determining countries' energy choices. In many parts of the developing world, simply providing access to energy remains the fundamental challenge. Globally, over three billion people still lack access to modern forms of energy. A recent report from the World Bank warns that, under current projections, the world is likely to fall short of reaching universal electrification by 2030, a key component of the UN *Sustainable Energy for All* goals.[22]

Many countries have abundant domestic reserves of "cheap" coal that could help meet growing energy needs. Yet, accounting for the public health and environmental costs makes it increasingly clear that coal dependency is imposing a huge and costly burden on societies. Research shows that in China, for example, coal combustion is the single biggest contributor to air pollution and was responsible for an estimated 366 000 premature deaths in 2013.[23] Another study showed that pollution from coal has contributed to shortening of life expectancy of 5.52 years on average among residents in Northern China, relative to those in the South. This difference was due to elevated rates of cardiorespiratory mortality that was an unintended consequence of a government policy to provide free coal for winter heating in the North.[24] Studies show that in India as many as 115 000 people *per* year die from pollution from coal-fired power plants.[25] Heavy smog from coal and other forms of fossil fuel pollution has become a highly visible indicator of poor air quality in major cities like Beijing and New Delhi, resulting in growing public pressure for policymakers to take action.

These types of near-term health considerations have moved the Chinese government to act to curtail coal use. They are also a major reason for US air and water pollution regulations aimed at coal-fired power plants (see Section 8).

China began piloting carbon cap-and-trade programs at the provincial level in 2013 and is expected to launch a national program in 2017 that will cover eight industries to begin with, including the power sector.[26,27] India has had a carbon tax in place since 2010, imposed on coal, lignite and peat. The so-called *Clean Environmental Cess* began at Rupees 50 *per* tonne and has been raised over time to Rupees 400 *per* tonne in 2016. The 2016 levy would increase the price of coal by approximately 20%. For the 2016–2017 fiscal year, the Government is expected to collect approximately $3.7 billion

(Rupees 23 944.4 Crore, where one crore equals 10 million Rupees) through this mechanism.[28] Since 2010, approximately \$1.8 billion (Rs. 12 430 Crore) from the coal tax revenue has been allocated to a National Clean Energy Fund.[29] India has set a goal of reaching 175 GW of renewable energy by 2022, with a National Solar Mission goal of 100 GW of solar by that time.[30]

In the US and Germany, efforts to implement national cap-and-trade or carbon-tax policies have foundered thus far. The closest the US has come was the 2009 American Clean Energy and Security Act, aimed at establishing a national cap-and-trade program, which passed the House but was never brought to a vote in the Senate due to insufficient support. There have been several proposals since then but none that have attracted significant political support. A recent proposal from a conservative group, the Climate Leadership Council, is garnering some interest, although it is unclear if it will have any real traction in Congress or with the Trump administration.[31] Germany's attempt to pass a coal tax failed in 2015 after bitter opposition from utility companies and unions.[32]

5   Opposition to Coal Regulation

Policies to regulate coal use or production face stiff resistance from coal companies and utilities that often have significant political clout. There is no doubt that across the world this has been a major factor in slowing down the transition away from coal to cleaner generating resources. Typical strategies include claims that curtailing coal use will cause the price of electricity to sky rocket, hurting consumers, or that the reliability of the electricity system will be threatened. In the US coal companies have also joined with other fossil fuel producers in mounting expensive misinformation campaigns aimed at denying the reality of climate change or public health costs of coal use. Coal company bankruptcies are on the rise in the US, which could mean that over time their political clout will also diminish.

Understandably, coal miners' and utility workers' unions have also resisted the regulation of coal because of concerns about associated job losses. Coal jobs have been declining over decades in many places because of increased mechanization of mining, but there are heightened concerns associated with rapid shifts that may be precipitated by a combination of cheaper alternatives to coal (natural gas, renewable energy) and climate policies. The reality is that coal mining is one of the most dangerous jobs in the world and coal companies have not always had the best interests of miners in mind. In the US coal companies have engaged in efforts to undermine safety regulations and cut back on pension and health benefits for miners.

Countries will have to find ways to address the legitimate economic concerns of workers displaced by the shift away from coal, otherwise coal communities will face significant harms (see Section 6). The political backlash from inequitable outcomes can also create a deeply polarized electorate and pose long-term challenges to socioeconomic progress.

6 Transition Assistance

In countries that are major coal producers, a shift away from coal to meet public health and climate goals raises the prospect of significant social and economic disruptions for coal miners and coal-dependent communities.

These types of shift are not new. The UK and Germany, among others, have already experienced these painful upheavals. The UK mining industry reached its heyday in the middle of the last century and declined steeply in the 1970s and 1980s, a period marked by mining job losses, bitter labor disputes and strikes. Germany, too, has seen coal mining decline through the 1960s to 1980s, which hit the Ruhr and the Saar regions especially hard. The mining of Hard coal (or Anthracite coal) is expected come to a stop in 2018 when subsidies are ended; however, the mining of lignite, or "brown coal," may continue. Reclamation of coal mines will cost 220 Euros *per* year indefinitely, paid for through an industry fund.[33]

Coal mining jobs have also been declining in the US for decades, primarily due to increased mechanization of the mining industry. In 2015, there were 65 971 employees in coal mines in the US Despite heated political rhetoric about the "war on coal," the reality is that even in the heart of coal country, in places like West Virginia, coal's contribution to local economies has been modest and declining for decades. With low natural gas prices putting further pressure on coal and ongoing coal-fired power plant retirements, the future of the coal mining industry looks grim.

China is in the midst of a major dislocation of coal miners as it seeks to shut down mines and curtail production. The government has promised to "reallocate" millions of workers employed in coal-dependent industries and find new jobs for them, budgeting the equivalent of $15 billion (100 billion yuan) over two years for these efforts.[34]

Job losses and economic transitions can arise for many reasons, including structural changes in an economy or trade, or changing technologies. How well workers and communities weather these changes comes down to transition assistance packages, efforts to diversify local economies, investments in worker retraining and education, and broader social safety-net programs. What is clear is that the latest round of changes, triggered by a shift away from coal to lower carbon energy sources, is likely to bring unprecedented changes to economies and their workforces. Some of these changes are positive: consider, for example, the huge and growing employment in the renewable energy sector. But some are more difficult. To successfully navigate these shifts and take into account the human dimensions of decarbonizing their economies, countries will have to integrate equity considerations into their climate and clean energy policies.

7 Carbon Capture and Storage: The Future of Coal?

Carbon capture and storage (CCS) is a technology innovation that would enable the CO_2 emissions from coal combustion in power plants or industrial facilities to be captured and stored underground indefinitely, thereby preventing its release into the atmosphere. Components of the technology have been in use for purposes such as enhanced oil recovery and a few demonstration projects are in operation around the world. But it is a long way from being widely deployed commercially on a large scale. If coal with CCS (or, for that matter, natural gas with CCS) is to be an important part of climate solutions, significant economic and technical hurdles will have to be overcome.

The concept of "clean coal" is an oxymoron because of all the public health and environmental harms caused across the fuel cycle of coal, including mining and transportation. However, a technology that would significantly cut its CO_2 profile would be a very significant improvement from a climate perspective. Hence there is strong interest from major coal producers to advance this technology through research and development (R&D) funding, tax credits, loans and other policies.

In the US, billions of dollars of federal aid have been invested in trying to develop CCS technology. The Bush administration provided millions of dollars in funding and loan guarantees for CCS, including for the FutureGen 1.0 project: a public-private partnership to build a power plant with Integrated Gasification Combined Cycle (IGCC) technology. That project was cancelled due to cost overruns. The 2009 American Recovery and Reinvestment Act (AARA) provided $3.4 billion for CCS programs, including industrial projects, the Clean Coal Power Initiative, and $1 billion for FutureGen 2.0, intended to be a clean coal repowering program and carbon dioxide (CO_2) storage network. In 2015, the FutureGen 2.0 project was also scrapped because of cost concerns and lack of technological progress.

The US Department of Energy (DOE) has maintained long-standing programs on Clean Coal research as well as industrial carbon capture and storage, through which it provides support and works in partnership with private companies. According to the DOE, the United States could store at least 2400 billion metric tons of carbon dioxide (CO_2) in saline formations, oil and gas reservoirs, and unmineable coal seams.[35]

In January 2017, the Petra Nova CCS project, located near Houston, Texas, came on line. A first-of-its-kind facility, designed to capture 90% of the total CO_2 emissions *via* post-combustion capture technology installed at an existing pulverized coal plant, it marks a significant milestone for commercial deployment of this technology. Another CCS project, the Kemper plant in Kemper County, Mississippi has suffered from repeated delays and cost overruns shouldered by ratepayers. The prospects of successful deployment of this plant any time soon seem dim at best.

China has a number of CCS demonstration projects at various stages of construction and development, a sign of strong government support for developing this technology. These include the Sinopec Shengli Power plant, Huaneng GreenGen IGCC Project, the China Resources Power (Haifeng) Integrated Carbon Capture and Sequestration Demonstration Project and the Shanxi International Energy Group Carbon Capture, Utilization and Storage (CCUS) Project.[36]

In a 2014 Joint Announcement on Climate Change, China and the US launched the US-China Clean Energy Research Center, to facilitate collaborative work in carbon capture and storage technologies.[37] China also participates in a number of other international research collaborations on CCS, including the EU-UK CCS Cooperative Action within China (COACH program), the China-EU Cooperation on Near Zero Emissions Coal (NZEC) and the Asia-Pacific Partnership on Clean Development and Climate.[38,39]

See also Chapter 7 in this book for further details on CCS.

8   Coal Regulation in the United States: A Political Football

Coal regulation in the United States illustrates two aspects of the American political system. First, Democrats and Republicans are deeply divided over coal regulation and so coal regulation is like an American "football" game, in which one political party advances down the field of regulation, followed by the other gaining possession of the ball and advancing in the other direction. Second, coal regulation is a good example of how legal authority is divided at the federal level between executive, legislative, and judicial branches, and further bifurcated between the federal government and states. All of these different actors play important roles in the regulation of coal, and this makes coal regulation a complex subject. The following examples illustrate these themes.

9   Regulating CO_2 Emissions from Coal: The Embattled "Clean Power Plan"

Coal-burning power plants are one of the largest single sources of carbon dioxide emissions in the United States, accounting for approximately 26% of all energy-related total emissions.[40] While coal-powered plants are declining precipitously due to their age and competition, falling prices for natural gas and renewable sources, they are still expected to account for approximately 20% of generation through 2030.[41] Thus, regulating CO_2 emissions from coal-fired power plants is and will remain a crucial component for the United States to lower its carbon emissions and meet its Paris Climate Agreement pledge to cut overall greenhouse gas emissions by 26–28% below 2005 levels by 2025.[42,43]

9.1   The Legal Route to Federal Regulation

Prior to the issuance of President Obama's Clean Power Plan (discussed in detail in the next paragraph), there was no direct federal regulation of CO_2 emissions from power plants. In fact, the George W. Bush administration had claimed that it had no authority to regulate CO_2 emissions, a claim that the United States Supreme Court rebuffed in a case brought by the Commonwealth of Massachusetts against the US Environmental Protection Agency (EPA). In that case, the court ruled that CO_2 and other greenhouse gases were "pollutants" within the meaning of the 1970 Clean Air Act, and therefore EPA had a duty to regulate them if it found that these pollutants presented a danger to public health, safety and the environment.[44]

In 2009, under President Obama, the EPA issued the requisite "endangerment" finding that emission of greenhouse gases posed a danger to public health and welfare.[44] This finding both enabled the EPA and obligated the EPA to develop regulations to limit CO_2 emissions from many sources, including coal plants. However, for several years, this authority remained largely dormant, as the Obama Administration and allies in congress tried to enact national climate change legislation rather than use the 1970 Clean Air Act,[45] which focused on more conventional and localized sources of pollutants.[46] When that legislative effort failed, President Obama decided to use the full range of his authority under the Clean Air Act. President Obama issued a "Climate Action Plan" that identified numerous pathways for federal agencies to reduce greenhouse gas emissions, including for the EPA to prescribe limits on carbon pollution from power plants.[47]

The Clean Air Act has no obvious or straightforward provision authorizing carbon pollution regulation. The Act is principally focused on five "criteria" pollutants: sulfur dioxide, nitrogen oxides, ozone, particulate matter and carbon monoxide, and authorizes the EPA to establish national health-based standards on these pollutants, and requires states to issue "state implementation plans" to meet these limits. The Act also empowers the EPA to establish source-by-source limits on emissions of these pollutants. The Act also specifically empowers the EPA to regulate "hazardous air pollutants" such as mercury, lead, and dioxins.[48]

However, the Act has a "catch-all" provision referred to as Section 111(d).[48] That section essentially provides that if a pollutant is not regulated under the specific sections addressing criteria pollutant or hazardous pollutants, that pollutant or source may be regulated under section 111(d). This section authorizes the EPA to establish standards of performance for sources that emit such pollutants, based on the "best system of emissions reduction" (BSER), and requires states to prepare plans to meet those standards of performance.[51] This catch-all provision has rarely been used—only five sources have been regulated under it—and it has never been tested in court.[49] Until now.

9.2   The Clean Power Plan

In 2014, using section 111(d) as authority, the EPA issued in draft form the nation's first-ever limits on carbon pollution from power plants in two separate rules, one for new plants, and one for existing plants. The EPA held a lengthy public comment on these rules, receiving over 4.3 million comments, the most ever received.[50] Most of the attention was directed towards the rule for existing plants. The coal industry, coal-dependent states and business organizations fiercely opposed the draft rule, while environmental NGOs, renewable energy developers and states with more-limited coal assets supported it with equal fervency.

The final rules were issued in October 2015. These require new, reconstructed or substantially modified coal-fired power plants to emit no greater than 1400 lb CO_2 per MWh.[51] The EPA based this limit on the premise that new coal-fired power plants could be equipped with carbon capture and storage devices (CCS) that would be capable of achieving these limits. This rule is currently under appeal by state and industry opponents, who claim that the EPA should not have relied upon CCS to establish this performance standard, as it is not technically or economically feasible to employ it at a large scale.[52] The outcome of this litigation may not prove to be consequential, as there are virtually no new coal-fired power plants being proposed in the United States due to the high cost of coal *versus* natural gas and renewable sources, conventional air pollution control requirements and general antipathy towards developing new high-carbon sources of power.

The consequential fight is over the rule that addresses existing coal (and gas-fired) power plants. In this rule, the EPA established a standard of performance that is calculated to reduce carbon pollution from existing power plants by approximately 32% below 2005 levels by 2025.[54] To put this standard in perspective, carbon emissions from power plants have already decreased by approximately 12% since 2005;[53] thus, the rule "codifies" the trajectory that the United States already is on.

The relatively modest reductions derive from the legal standard. Under this section of the Clean Air Act, the EPA is required to set standards based on the "best system of emissions reduction" (BSER), which "has been adequately demonstrated".[48] The EPA interpreted this language to preclude setting standards that would force new technology development or cause a widespread upheaval of the existing electric grid.

The EPA determined that the "best system of emissions reduction" for reducing CO_2 emissions is three approaches already in place, which EPA referred to as the three "building blocks":

- Improving the heat rate efficiency of coal fired plants
- Switching from coal to natural gas
- Substituting low- or no-carbon generation (such as wind and solar) for coal and gas generation.[54]

The EPA then quantified the required emissions reduction for both coal and gas-fired plants using these three approaches. Dividing the nation into three existing electric grids, the EPA determined the percentage reductions that were possible in each grid using the above approaches. The EPA then selected as a national standard the reductions from the lowest reducing grid, and translated that into CO_2 standards for coal plants (1305 lb CO_2 per MWh) and gas plants (771 lb CO_2 per MWh).[60] Importantly, the rule does not require existing plants to meet these rates; rather, each state must achieve these overall rates through their specific mix of energy generation.

The EPA then translated these rates into state specific targets that reflect states' existing mix of coal, gas and other sources. The targets are expressed as both rate-based goals (pounds of CO_2 *per* MMBTU), or mass-based targets (overall aggregate limit on CO_2). The rule gives states the choice of complying either with the rate-based targets or the mass-based targets.[60] And states are not limited to using these three approaches to meeting the targets. For example, they may invest in energy efficiency measures to lower overall demand so as to meet the mass-based limit. In addition, the EPA allows for states, or power generators, to trade "emission credits" with one another, and established guidelines for this trading. For example, a coal-dependent state or utility may find it difficult to lower emissions on their own to meet the targets; that state, or the individual coal-fired plants, are allowed to offset their emissions by purchasing emission credits from, *e.g.* a power generator using wind or solar energy.

The rule requires states to submit state implementation plans by 2018, meet certain interim reduction milestones by 2021, and fully achieve the reduction targets by 2030.[54]

The rule is widely considered to be a key component of the United States' pledge to achieve a 26–28% reduction from 2005 levels by 2025.[55]

9.3   The Litigation

After the final rule was published, the rule's opponents (26 states, numerous coal companies, utilities and manufacturers) filed suit in the Court of Appeals for the District of Columbia (D.C. Circuit). The first round in the litigation was over whether the court would put the rule on hold while it was being litigated. Ordinarily, when a litigant challenges a regulation, the mere filing of the suit does not stay the regulation; in order to get a stay, opponents must prove that they are likely prevail and will suffer irreparable harm if forced to comply with the regulation while the appeal is pending. The argument for "irreparable harm" seemed particularly challenging here, given that power plants were given until 2030 to come into compliance with rule, and the litigation would take approximately two to three years to be resolved.

The D.C. Circuit rejected issuing a stay, but expedited the briefing and hearing process, perhaps reasoning that a prompt decision on such an important case would be in everyone's interest.[56] The opponents then asked the United States Supreme Court to issue a stay. By a 5 : 4 vote, the court did so[57]—the first time in history that the Supreme Court issued a stay on a regulation pending in the D.C. Circuit. Three days later, one of the justices joining the majority, Antonin Scalia, passed away.

In September 2016, the D.C. Circuit Court of Appeals heard the case; a decision is pending as of the time of this writing.

Ultimately, the legal arguments boil down to an age-old debate about the proper balance between legislative and executive authority and, in particular, how explicit congressional authorization needs to be for an agency such as the EPA to act. Under American law, agencies do not have any inherent authority; all of their powers come explicitly or implicitly from laws enacted by congress.

The opponents argue that a rule of this magnitude should be explicitly authorized by congress. The seldom-used, largely dormant catch-all provision of section 111(d) is silent on greenhouse gases and does not explicitly authorize EPA to issue a national cap on carbon emissions from power plants and steer the economy towards cleaner and renewable energy sources. Indeed, such a legislative provision was attempted in congress a few years before the rule, and failed. The opponents cite the colorful language of now-deceased Justice Scalia: congress does not "hide an elephant in a mouse-hole", and courts should "greet ... with a measure of skepticism" claims by the EPA to have "discover[ed] in a long-extant statute an unheralded power to regulate a significant portion of the American economy" and make "decisions of vast economic and political significance."[58]

On the other hand, the proponents argue that the Supreme Court has twice ruled that the EPA does have the power to regulate greenhouse gases under the Clean Air Act, that carbon emissions from power plants are the largest stationary source of such emissions, and that congress enacted section 111(d) and used broad and elastic terms such as "best system of emissions reduction" to give the EPA the flexibility to evolve its regulations to address pollutants as they become of concern over time. To the proponents, as long as the statute does not explicitly *foreclose* the chosen approach, it is within the EPA's discretion to employ it.

9.4   Trump Administration

While the case is pending, the political football of coal regulation has changed hands. Newly elected President Trump has vowed to repeal the Clean Power Plan and has chosen as the head of EPA a lawyer (Scott Pruitt, former Attorney General of Oklahoma) who led the litigation against it. And in March 2017, President Trump issued an executive order directing the EPA to repeal the rule.[59]

However, the Plan cannot be eviscerated with the stroke of a pen. The Trump Administration will have to follow the same process to repeal the rule as the Obama Administration did to issue it: perform technical analysis, issue a draft rule, take comments upon it, and issue a final rule subject to court challenge. This process will take years.

The options facing the Trump administration are to: (1) repeal it entirely; (2) leave some portions of it in place; or (3) repeal it, but replace it with an alternative. Option 1 is a legally risky approach, because the Supreme Court held that the EPA has a legal obligation to regulate greenhouse gases if it finds that these gases cause an endangerment, and the EPA made the endangerment finding in 2009. It is possible that the Trump administration could try to escape this quandary by rescinding the "endangerment" finding itself, but this would fly in the face of an overwhelming consensus of scientists employed by the federal government and elsewhere, and it is highly likely that a court would negate such a fundamental switch in position.

Option 2 is a safer approach. Under Option 2, the Trump administration might jettison the portions of the rule that are based on substituting gas and renewables for coal, and leave in place the part of the rule that requires coal plants to operate more efficiently. This would drastically reduce the expected CO_2 emission reductions but would allow the Trump administration to claim that they are not ignoring the problem altogether.

Option 3, at this time, seems unlikely. Candidate Trump called global warming a hoax, and the Trump administration seems determined to cut regulations, not issue new ones. There is some bi-partisan support for a carbon tax to replace direct regulation, but this far no Republican leaders in congress have embraced that proposal.

Further complicating the matter is that the courts have not ruled on whether the Plan is lawful. It is possible that the court may opt not to rule, and send it back to the EPA. If this occurs, the EPA under its new administrator, Scott Pruitt, would have considerable latitude. However, the court may decide to rule, as the issue of the EPA's authority under this section of the Clean Air Act has not been resolved and may be important to inform future efforts. If the court were to rule, *and* uphold the Plan, this could tie the Trump Administration's hands, as it would be much harder to justify repealing a Plan that has been found to be lawful.

As the reader can see, the regulation of CO_2 at the federal level does not follow a logical path. Right now, Republican leaders have possession of the football, but it is a long game and one in which the umpires (the federal courts) also have an important say in the outcome.

When, and even whether, CO_2 emissions from coal plants will be controlled by federal policy remains very uncertain.

10   Regional/State Regulation of CO_2 Emissions from Coal Fired Plants

In the United States, it is often the case that state regulation precedes federal regulation and is more stringent. In large part, this is because in many states there is less partisan division than at the federal level. This is true of the CO_2 regulation of coal plants: emissions limits were initially pioneered by states and, in many instances, are stricter than the Clean Power Plan targets.

One of the country's first serious efforts to curb CO_2 emissions came from the Regional Greenhouse Gas Initiative (RGGI) in 2008.[60] RGGI is a CO_2 cap-and-trade program for the electric sector with participation from nine northeastern/mid-Atlantic states. It requires coal and gas-fired power plants of 25 megawatts or more to purchase "allowances" to emit CO_2, the vast majority of which are auctioned off at quarterly auctions. These allowances can then be traded in a secondary market. The RGGI program has a minimum allowance price, a "cost containment reserve" to guard against allowance price volatility, and provisions to allow offsets should allowances become too scarce.

Like the European Emissions Trading System, the RGGI program was initially hampered by an overabundance of allowances. RGGI did not anticipate the precipitous decline in natural gas prices due to hydraulic fracturing, and the rapid fuel switching from coal to gas which lowered emissions and the demand for allowances. However, the RGGI states took decisive action in 2013 to cut the cap in half to actual 2014 emissions levels, and to reduce it by 2.5% thereafter.[61] The RGGI states also agreed to lower the cap to deplete the large privately held supply of allowances that had accumulated during the years of low allowance prices. The RGGI allowance prices, which were less than $2 per ton in 2013, are now steadily rising, though there have been ups and downs on the allowance pricing.

Since 2008, RGGI auction revenues have generated over a billion dollars, and most of this revenue has been re-invested in energy efficiency and renewable energy.[62] Independent economic analyses in 2012 and 2015 have determined that this "auction and invest" approach has brought over $3 billion in economic value to this region.[63,72]

Due to the initial abundance of allowances, the RGGI program played more of an indirect role in lowering emissions, primarily by providing ample funds for states to decrease overall electric demand through efficiency programs. However, as the cap tightens and the private bank of allowances is depleted, it is expected that the cap itself will constrain emissions. Current projections are that CO_2 emissions from the RGGI state power plants are expected to be half of 2005 levels by 2020. This is a significantly steeper reduction than is called for by the Clean Power Plan.

At the same time that the RGGI program was forged by eastern states, the Western Climate Initiative was formed in 2008 by nine American states and three Canadian provinces to establish a linked, all-economy-wide carbon-trading mechanism.[64] While several of these states and provinces did not push forward on this initiative, California and Quebec did. They each developed their own economy-wide cap-and-trade systems and in 2014 formally linked their programs together. The joint California–Quebec program covers all the greenhouse gases under the Kyoto protocol. It establishes an overall cap on emissions and requires emitters of more than 25 000 metric tons CO_2 equivalents to purchase allowances. As of 2015, the program applies to electric generators, industrial sources, distributors of transportation and other fuels, and distributors of natural gas, covering approximately 87% and 77% of California's and Quebec's emissions, respectively. The cap will decrease by approximately 3% *per* year in CA and 4% *per* year in Quebec, steadily driving down emissions from these sectors.[65] The program also has a minimum floor price that starts at $10 per ton, rising annually by 5% plus the rate of inflation. The market clearing price at the latest auction (February 2016) was $12.73.[66] Thus far, allowance auctions have generated approximately $4 billion[76] in funds that have been deposited into California's

Greenhouse Gas Reduction Fund, and over \$1 billion[67] into Quebec's Green Fund.

In 2017, Ontario commenced implementing a cap-and-trade program. Ontario is a member of the Western Climate Initiative (WCI), and it used that model to design its program. It is widely expected that once it is up and running it will link to the joint California–Quebec program.[68]

In the absence of national climate policy in the United States, there is increasing interest in linking these two regional programs. Were Ontario to join the California–Quebec program, and the WCI and RGGI programs to formally link, this would constitute an international carbon-trading market of nearly 103 million people (about a third of the US and Canadian population) and a GDP of \$6.2 trillion (the fourth largest economy of the world).

The premise of cap-and-trade is that multiple actors can reduce emissions more cost-effectively if they cooperate than if they try to do it alone. Under a cap-and-trade program, an emitter with low cost-reduction opportunities can sell allowances to an emitter with higher reduction costs, who will pay for allowances as long as the price is lower than their own cost of reduction. The logic of cap-and-trade suggests that the larger the market, the greater the opportunity for these cost-reducing trades. In addition, a larger market is more likely to have stable and predictable prices as exogenous variables will have less of an overall effect.

This potential for cost savings through a bigger trading market will be very tempting, particularly for sub-nationals with individual or regional climate goals that will require sharp emissions reductions in the near- and long-term.

For example, numerous states in Mexico and the United States, and provinces in Canada, have entered into an "under2Mou" in which they have agreed to lower their respective emissions by 80–95% below 2005 levels by 2050.[69] The signers include eleven states in Mexico, ten in the United States (including four RGGI states), and five provinces in Canada (including Quebec and Ontario).

In addition, many of the states and provinces that are in cap-and-trade programs have established very ambitious goals for 2030. For example, the New England Governors/Eastern Canadian Premiers have collectively pledged to lower emissions in their region by 35–45% below 2005 levels by 2030. California has enunciated three extremely ambitious goals for 2030 (50% renewable energy, 50% drop in oil use, and doubling energy efficiency). Ontario has a 2030 target of a 37% reduction.

A broad, integrated North American trading market could make it far easier for these states and provinces to meet their goals. This seems to be the thinking of Governor Cuomo, who announced last fall that he had instructed his staff to explore connecting the RGGI to the California–Quebec program. As he stated, "Connecting these markets would be more cost-effective and stable, thereby supporting clean energy and driving international carbon emission reductions."[70]

Further enhancing the possibility of linkage is that the programs share some fundamental features. Both programs largely auction off, rather than give away, allowances. Both programs have components such as declining caps, minimum floor prices, cost containment reserves, and market monitoring to prevent market manipulation. Both programs allow for purchases of similar categories of offsets if allowance prices reach certain levels.

However, there are some important differences. RGGI covers only CO_2 and only the electric sector as a source. RGGI allowance prices are currently about half of those of the joint California–Quebec cap-and-trade program. However, there are reasons to be hopeful that the two programs' pricings will converge over time. The RGGI allowance prices are expected to rise as the cap tightens and the bank of allowances is used up. But, if as expected, one sees the RGGI allowance prices steadily rising and moving toward convergence, it seems likely that linkage talks may intensify, and it would not be unreasonable to predict that the programs link together in the early 2020s. This linked international carbon market would be a tremendous symbol of progress and serve as a template for North American carbon pricing.

11 Regulation of Mercury and Other Air Toxics from Coal Plants: What a Long, Strange Trip It's Been

In addition to CO_2 emissions, coal-fired power plants are a major source of mercury emissions.[71] Regulating these emissions has taken twenty years and, like CO_2 regulation, has been a political football passed back and forth between the legislative, executive and judicial branches of the federal government.

The path to regulation began in 1990, when Congress amended the Clean Air Act to add numerous new provisions, including sections mandating that the EPA tackle approximately 180 hazardous air pollutants and establishing a cap-and-trade program to reduce acid rain.[72] Mindful of that the fact that pollution controls on coal plants under the acid-rain provisions might also reduce hazardous air emissions, congress enacted unique provisions for the regulation of air toxics from coal plants.[84] It ordered the EPA to conduct a study of the public health impacts of mercury emissions in light of the reductions that would occur under other programs. If the EPA determined that it would be "necessary and appropriate" to order additional pollution controls on coal plants under the air toxics program, it could do so.[84]

In 1998, the EPA completed the study and issued a finding in 2000 that it was "necessary and appropriate" to regulate coal plants under this section.[73] This finding obligated the EPA to issue limits by 2004.

By that time, President George W. Bush replaced President Clinton and the EPA rescinded its "appropriate and necessary" finding, over the recommendations of its own scientific advisory council. Rather than issuing regulations under the air-toxics program, which would have required each individual facility to use maximum achievable control technology (MACT), the EPA instead proposed a Clean Air Mercury Rule (CAMR), which established a total cap on mercury emissions and a voluntary cap-and-trade to implement it.[74]

In 2008, the D.C. Circuit court vacated the Bush administration's decision not to regulate toxics from coal plants, and the CAMR.[74]

Political power again switched hands (from George W. Bush to Barack Obama). The EPA re-issued the 2000 "appropriate and necessary" finding and also issued regulations limiting mercury emissions from approximately 600 power plants.[75] The emission limits are based on what is achievable using "maximum achievable control technology." For new sources, MACT is the emissions achieved by the best performing similar source. For existing sources, MACT requires emissions to be as stringent as the reductions achieved by the average of the top 12% lowest emitting sources.[83]

The EPA estimated that the rule would cut mercury emissions by about 90% and also substantially reduce emissions of acid gases and other pollutants such as sulfur dioxide.[83] When this rule was issued, approximately 40% of the nation's coal-fired power plants had not installed the requisite controls.[83]

Industry opponents immediately challenged this regulation in court. They fired a large array of objections, and one stuck. It turns out that when the EPA issued the "appropriate and necessary" finding it did not explicitly factor in the cost of compliance.[76] Rather, the EPA reasoned that the words "appropriate and necessary" required it to examine whether mercury emissions from coal plants continued to threaten public health, whether controls were available, and whether the other Clean Air Act programs had effectively addressed the threat.[77] The EPA further reasoned that it would take costs into effect when setting the overall level of protection. And the EPA did so, preparing a cost–benefit analysis when it issued the rule, finding that compliance would cost industry approximately $9.6 billion *per* year, would provide direct health benefits of $4–6 million *per* year and indirect "co-benefits" of reducing other pollutants worth approximately $37–90 billion *per* year.[77]

With another 5 : 4 majority (all Republican appointees in the majority; all Democratic appointments in the minority), the court ruled that the EPA was required to consider costs in the threshold decision of whether to regulate mercury emissions and could not defer that consideration to a later phase in the process.[78]

While this court decision seemed to have struck a significant blow, it turns out to be a Pyrrhic victory. In April 2016, the EPA re-issued its "appropriate and necessary" finding, this time complete with cost and benefit calculations.[79] The rule remains in effect at the time of writing.

However, this supplemental finding is being challenged by industry opponents, who are claiming that the cost–benefit analysis is flawed because a significant percentage of the benefits from the rule come from reductions in pollutants other than mercury.[80] So far, the courts have not been persuaded by this argument and have refused to enter a stay on the rule.

However, the new EPA Administrator, Scott Pruitt, sued the EPA to block this rule when he was the attorney general of Oklahoma. It is possible that he may seek to repeal or weaken the rule, but no announcements have been made to this effect at the time of this writing. In any event, many coal plants have already installed the requisite pollution control and many coal plants are expected to retire in the next coming years, due to plunging prices for natural gas and renewables.

12 Acid Rain: A Notable Success Story

One of the biggest impacts of coal-burning power plants, particularly in the northeast, is "acid rain," or highly sulfurous precipitation that has wreaked havoc on the forests and water resources of the eastern United States.[81] In 1990, a bi-partisan congress, with the support of President George H. W. Bush, devised what was then a novel approach to the problem. Rather than employ traditional "command and control" regulations from the EPA, congress directed the EPA to establish a "cap-and-trade" program on SO_2 emissions, the primary precursor of acid rain.[82] Phase I (1995–1999) applied to over 200 coal-fired power plants, mostly in the east and mid-west. Phase II, which began in 2000, covered nearly all fossil-fuel plants in the US and limited overall emissions to 8.95 million tons, about half of 1980 levels.[83] The government gave permits to emit, called allowances, denominated in tons of SO_2 emissions, to power plants covered by the law. If annual emissions at a regulated facility exceeded the allowances allocated to that facility, the owner could buy allowances or reduce emissions, either by installing pollution controls, changing the mix of fuels used to operate the facility, or by scaling back operations. If emissions at a regulated facility were reduced below its allowance allocation, the facility owner could sell the extra allowances or bank them for future use. These opportunities created incentives to find the least-cost method to reduce pollution.[84]

Under this program, the government gave "allowances" to these sources, which they would need to emit SO_2. The sources may use the allowances to cover their own SO_2 emissions, sell allowances they didn't need, or buy additional allowances. In essence, the program combines perhaps the best of government and the best of markets, as government sets the overall limit based on public policy considerations and the market determines the lowest cost means to meet that limit.[83]

This program has been hailed as the "greatest green success story of the past decade"[85] and was a role model for the RGGI CO_2 program as well as other carbon trading programs such as the European Trading System (ETS).[86] The program succeeded in cutting SO_2 emissions in half, and did so three years ahead of schedule and at a much lower than predicted cost.[86]

13 Conclusion

The regulation of coal is one of the most important policy drivers in meeting our global obligations under the Paris Agreement. As this chapter illustrates, major countries are using a variety of tools, including carbon pricing, incentives, research and development funding, and direct emissions regulation to lower carbon dioxide and other pollutant emissions from the combustion of coal. While some countries (*e.g.* China and Germany) offer stable political environments for these policies, others (*e.g.* the United States) do not. However, market forces and sub-national policies in the United States and, in particular, the plentiful supplies of natural gas as well as wind and solar resources, will likely continue to contribute to the downward trend of coal-related emissions even in the absence of a stable, federal policy.

References

1. United Nations Framework Convention on Climate Change (UNFCCC). 2015. Paris Agreement. http://unfccc.int/paris_agreement/items/9485. php. Accessed 17 Apr 2017.
2. International Energy Agency (IEA). 2016. CO_2 emissions from fuel combustion: 2016 edition. http://www.iea.org/bookshop/729-CO2_ Emissions_from_Fuel_Combustion. Accessed 8 Apr 2017.
3. International Energy Agency (IEA). 2016. Medium-Term Coal Market Report 2016: Market Analysis and Forecasts to 2021. https://www.iea.org/ newsroom/news/2016/december/medium-term-coal-market-report-2016. html, Accessed 8 Apr 2017.
4. U.S. Environmental Protection Agency (EPA). No date. https://www.epa. gov/mats/cleaner-power-plants. Accessed 8 Apr 2017.
5. U.S. Environmental Protection Agency (EPA). No date. https://www.epa. gov/airmarkets. Accessed 8 Apr 2017.
6. International Energy Agency (IEA). 2016. Key Coal Trends. http://www. iea.org/publications/freepublications/publication/KeyCoalTrends.pdf, Accessed 8 Apr 2017.
7. National Bureau of Statistics of China. 2017. Statistical Communiqué of the People's Republic of China on the 2016 National Economic and Social Development. http://www.stats.gov.cn/english/PressRelease/ 201702/t20170228_1467503.html. Accessed 8 Apr 2017.
8. U.S. Energy Information Administration (EIA). 2016. Annual Coal Report. https://www.eia.gov/coal/annual/. Accessed 8 Apr 2017.
9. U.S. Department of State. 2015. U.S.A. First INDC Submission. http:// www4.unfccc.int/ndcregistry/PublishedDocuments/United%20States% 20of%20America%20First/U.S.A.%20First%20NDC%20Submission.pdf. Accessed 10 Apr 2017.
10. U.S. Environmental Protection Agency (EPA). 2017. Draft Inventory of U.S. Greenhouse Gas Emissions and Sinks: 1990-2015. https://www.epa. gov/ghgemissions/inventory-us-greenhouse-gas-emissions-and-sinks. Accessed 10 Apr 2017.
11. People's Republic of China National Development and Reform Commission. 2015. PRC First INDC Submission. http://www4.unfccc.int/ ndcregistry/PublishedDocuments/China%20First/China%27s% 20First%20NDC%20Submission.pdf. Accessed 10 Apr 2017.

12. Greenpeace Asia. 2017. China forecasts fourth year of stable or declining CO2 emissions as world awaits Trump climate action – Greenpeace. Press release. http://www.greenpeace.org/international/en/press/releases/2017/China-forecasts-fourth-year-of-stable-or-declining-CO2-emissions-as-world-awaits-Trump-climate-action—Greenpeace/. Accessed 10 Apr 2017.

13. International Energy Agency (IEA). 2016. World Energy Outlook 2016. http://www.worldenergyoutlook.org/publications/weo-2016/. Accessed 10 Apr 2017.

14. M. Forsythe, 2017. China cancels 103 coal plants, mindful of smog and wasted capacity. *New York Times*, 18 January 2017. https://www.nytimes.com/2017/01/18/world/asia/china-coal-power-plants-pollution.html. Accessed 10 Apr 2017.

15. M. Forsythe, 2017. China aims to spend at least $360 billion on renewable energy by 2020. *New York Times*, 5 January 2017. https://www.nytimes.com/2017/01/05/world/asia/china-renewable-energy-investment.html. Accessed 10 Apr 2017.

16. K. Appunn, 2017. Germany's energy consumption and power mix in charts. Clean Energy Wire. https://www.cleanenergywire.org/factsheets/germanys-energy-consumption-and-power-mix-charts. Accessed 10 Apr 2017.

17. R. Martin, Germany runs up against the limits of renewable energy, *MIT Technol. Rev.*, 2016, https://www.technologyreview.com/s/601514/germany-runs-up-against-the-limits-of-renewables/. Accessed 10 April 2017.

18. Federal Ministry of Economic Affairs and Energy. 2016. Minister Gabriel: 2016 Renewable Energy Sources Act marks paradigm shift and launches next phase of energy transition. http://www.bmwi.de/Redaktion/EN/Pressemitteilungen/2016/20160608-gabriel-eeg-2016-schafft-paradigmenwechsel-und-ist-start-fuer-die-naechste-phase-der-energiewende.html. Accessed 10 Apr 2017.

19. S. Amelang, 2016. When will Germany finally ditch coal? Clean Energy Wire. https://www.cleanenergywire.org/factsheets/when-will-germany-finally-ditch-coal. Accessed 10 Apr 2017.

20. Federal Ministry for the Environment, Nature Conservancy, Building and Nuclear Safety. 2016. Climate action plan 2050. http://www.bmub.bund.de/en/topics/climate-energy/climate/details-climate/artikel/climate-action-plan-2050/?tx_ttnews%5BbackPid%5D=3915. Accessed 10 Apr 2017.

21. Eurelectric. 2017. European electricity sector gears up for the energy transition. http://www.eurelectric.org/media/318381/2017-04-05-eurelectric-press-release-on-energy-transition-statement-launch-of-cep-papers-embargo-9-am-542017.pdf. Accessed 10 Apr 2017.

22. World Bank. 2017. "Global Tracking Framework 2017: Progress Towards Sustainable Energy" (April), World Bank, Washington, DC., DOI: 10.1596/978-1-4648-1084-8. http://gtf.esmap.org/data/files/download-documents/eegp17-01_gtf_full_report_final_for_web_posting_0402.pdf. Accessed 10 Apr 2017.

23. GBD MAPS Working Group. 2016. Burden of disease attributable to coal-burning and other air pollution sources in China. https://www.healtheffects.org/publication/burden-disease-attributable-coal-burning-and-other-air-pollution-sources-china. Accessed 10 Apr 2017.

24. Y. Chen, A. Ebenstein, M. Greenstone and H. Li, Evidence on the impact of sustained exposure to air pollution on life expectancy from China's Huai River policy, *Proc. Natl. Acad. Sci. U. S. A.*, 2013, **110**(32), DOI: 1073/pnas.1300018110. Accessed 10 Apr 2017.

25. D. Goenka and S. Guttikunda 2013. Coal Kills: An assessment of death and disease caused by India's dirtiest energy source. A report by Conservation Action Trust and Greenpeace India. http://www.greenpeace.org/india/en/publications/Coal-Kills/. Accessed 10 Apr 2017.

26. J. Swartz, 2016. China's national emissions trading system: Implications for carbon markets and trade. ICTSD issue paper no. 6. http://www.ieta.org/resources/China/Chinas_National_ETS_Implications_for_Carbon_Markets_and_Trade_ICTSD_March2016_Jeff_Swartz.pdf. Accessed 10 Apr 2017.

27. L. He, 2017. China's national carbon trading rollout expected to have major impact on key industries. South China Morning Post, April 2, 2017. http://www.scmp.com/business/companies/article/2084151/chinas-national-carbon-trading-rollout-expected-have-major-impact. Accessed 10 Apr 2017.

28. Press Trust of India (PTI). 2016. Government may get Rs 23,944 crore from clean environment cess in FY17. *Economic Times*, August 11, 2016. http://economictimes.indiatimes.com/industry/banking/finance/government-may-get-rs-23944-crore-from-clean-environment-cess-in-fy17/articleshow/53651980.cms. Accessed 10 Apr 2017.

29. S. Mahapatra, 2017. $1.8 Billion of India's coal tax invested in renewable energy so far. https://cleantechnica.com/2017/02/21/1-8-billion-indias-coal-tax-invested-renewable-energy-far/, Accessed 10 Apr 2017.

30. Government of India, Ministry of New and Renewable Energy. National Solar Mission. http://www.mnre.gov.in/solar-mission/jnnsm/introduction-2/. Accessed 10 Apr 2017.

31. Climate Leadership Council. 2017. The conservative case for carbon dividends. https://www.clcouncil.org/wp-content/uploads/2017/02/TheConservativeCaseforCarbonDividends.pdf. Accessed 10 Apr 2017.

32. AFP. 2015. German government drops plans for contested coal tax. July 2, 2015. https://phys.org/news/2015-07-german-contested-coal-tax.html. Accessed 10 Apr 2017.

33. E. Marx, 2016. Can Germany ditch coal? The powerhouse of the European Union struggles to cut the dirtiest fossil fuel. ClimateWire, January 20, 2016. https://www.scientificamerican.com/article/can-germany-ditch-coal/. Accessed 10 Apr 2017.

34. D. Stanway, 2016. China allocates 100 billion yuan to deal with job losses from capacity cuts: ministry http://www.reuters.com/article/us-china-economy-overcapacity-idUSKCN0VY09Y. Accessed 10 Apr 2017.

35. U.S. Department of Energy, National Energy Technology Laboratory (NETL). The United States 2012 carbon utilization and storage atlas. https://energy.gov/fe/articles/does-carbon-utilization-and-storage-atlas-estimates-least-2400. Accessed 10 Apr 2017.

36. Global CCS Institute. Large scale CCS projects. https://www.globalccsinstitute.com/projects/large-scale-ccs-projects. Accessed 10 Apr 2017.

37. The White House, Office of the press secretary. 2014. U.S.-China Joint announcement on Climate Change. Nov 11, 2014. https://obamawhitehouse.archives.gov/the-press-office/2014/11/11/us-china-joint-announcement-climate-change. Accessed 10 Apr 2017.

38. D. Seligsohn, S. Forbes, Y. Liu, Z. Dongjie and L. West. 2010. CCS in China: Toward an environmental, health and safety regulatory framework. World Resources Institute. http://www.wri.org/publication/ccs-china. Accessed 10 Apr 2017.

39. Global CCS Institute. The global status of CCS: 2016 summary report. https://www.globalccsinstitute.com/publications/global-status-ccs-2016-summary-report, Accessed 10 Apr 2017.

40. U.S. Energy Information Administration FAQs, https://www.eia.gov/tools/faqs/faq.cfm?id=77&t=11. Accessed Mar 2017.

41. U.S. Energy Information Administration Annual Energy Outlook 2017, "Table: Electricity Generating Capacity", https://www.eia.gov/outlooks/aeo/data/browser/#/?id=9-AEO2017&region=0-0&cases=ref2017~ref_no_cpp&start=2015&end=2030&f=A&linechart=ref2017-d120816a.4-9-AEO2017~ref_no_cpp-d120816a.4-9-AEO2017&sourcekey=0. Accessed Mar 2017.

42. Massachusetts v. Environmental Protection Agency, No. 05-1120, U.S Court of Appeals for the District of Columbia Circuit, 549 US 497 https://www.supremecourt.gov/opinions/06pdf/05-1120.pdf. Accessed Mar 2017.

43. U.S Cover Note INDC and Accompanying Information, http://www4.unfccc.int/submissions/INDC/Published%20Documents/United%20States%20of%20America/1/U.S.%20Cover%20Note%20INDC%20and%20Accompanying%20Information.pdf. Accessed Mar 2017.

44. Federal Register, Vol. 74, No. 239, https://www3.epa.gov/climatechange/Downloads/endangerment/Federal_Register-EPA-HQ-OAR-2009-0171-Dec.15-09.pdf. Accessed Mar 2017.

45. Clean Air Act Amendments of 1970, 42 U.S.C §7401 et seq. (1970), http://uscode.house.gov/view.xhtml?req=(title:42%20section:7401%20edition:prelim)%20OR%20(granuleid:USC-prelim-title42-section7401)&f=treesort&edition=prelim&num=0&jumpTo=true#executivedocument-note. Accessed Apr 2017.

46. Summary of the Clean Air Act, https://www.epa.gov/laws-regulations/summary-clean-air-act. Accessed Mar 2017.

47. The President's Climate Action Plan, June 2013, p. 8, https://www.whitehouse.gov/sites/default/files/image/president27sclimateactionplan.pdf. Accessed Mar 2017.

48. Clean Air Act Standards of Performance for New Stationary Sources, 42 U.S.C. § 7411; https://www.law.cornell.edu/uscode/text/42/7411. Accessed Mar 2017.

49. See Phosphate Fertilizer Plants, Final Guideline Document Availability, 42 Fed. Reg. 12,022 (Mar. 1, 1977); Emission Guideline for Sulfuric Acid Mist, 42 Fed. Reg. 55,796 (Oct. 18, 1977); Kraft Pulp Mills; Final Guideline Document; Availability, 44 Fed. Reg. 29,828 (May 22, 1979); Primary Aluminum Plants; Availability of Final Guideline Document, 45 Fed. Reg. 26,294 (Apr. 17, 1980); Standards of Performance for New Stationary Sources and Guidelines for Control of Existing Sources: Municipal Solid Waste Landfills, 61 Fed. Reg. 9,905 (Mar. 12, 1996).

50. EPA Fact Sheet: Overview of the Clean Power Plan, https://www.epa.gov/cleanpowerplan/fact-sheet-overview-clean-power-plan. Accessed Mar 2017.

51. EPA Fact Sheet: Carbon Pollution Standards, https://www.epa.gov/sites/production/files/2015-11/documents/fs-cps-overview.pdf. Accessed Mar 2017.

52. State of North Dakota's Opening Brief, State of North Dakota v. U.S. Environmental Protection Agency, No. 15-1381, U.S Court of Appeals for the District of Columbia Circuit, 2017, https://www.edf.org/sites/default/files/content/2017.02.02_nd_final_opening_brief.pdf. Accessed Mar 2017.; Respondent EPA's Initial Brief, State of West Virginia v. U.S. Environmental Protection Agency, No. 15-1363, Document #1609995, U.S Court of Appeals for the District of Columbia Circuit, 2016, https://www.edf.org/sites/default/files/content/epa_final.pdf. Accessed Mar 2017.

53. Today in Energy, May 9, 2016; https://www.eia.gov/todayinenergy/detail.php?id=26152. Accessed Mar 2017.

54. Respondent EPA's Initial Brief, State of West Virginia v. U.S. Environmental Protection Agency, No. 15-1363, Document #1605911, U.S Court of Appeals for the District of Columbia Circuit, 2016, p. 12-13, https://www.edf.org/sites/default/files/content/epa_merits_brief_-_march_28_-_2016.pdf. Accessed Mar 2017; EPA Fact Sheet: Overview of the Clean Power Plan, https://www.epa.gov/cleanpowerplan/fact-sheet-overview-clean-power-plan. Accessed Mar 2017.

55. W. Cornwall, United States will miss Paris climate targets without further action, study finds, *Science Magazine*, 26 Sep 2016, http://www.sciencemag.org/news/2016/09/united-states-will-miss-paris-climate-targets-without-further-action-study-finds. Accessed Mar 2017.

56. Court Order, State of West Virginia v. U.S. Environmental Protection Agency, No. 15-1363, Document #1594951, U.S Court of Appeals for the District of Columbia Circuit, 2016, http://blogs.edf.org/climate411/files/2016/01/STAY-DENIAL.pdf. Accessed Mar 2017.

57. Court Order, Chamber of Commerce v. EPA, No. 15A787, Order List 577, Supreme Court of the United States, 2016, https://www.supremecourt.gov/orders/courtorders/020916zr3_hf5m.pdf. Accessed Mar 2017.

58. Utility Air Regulatory Group. v. Environmental Protection Agency (2014). No. 12-1146, https://www.supremecourt.gov/opinions/13pdf/12-1146_4g18.pdf. Accessed Mar 2017.

59. Presidential Executive Order on Promoting Energy Independence and Economic Growth, https://www.whitehouse.gov/the-press-office/2017/03/28/presidential-executive-order-promoting-energy-independence-and-economi-1. Accessed Mar 2017.

60. Regional Greenhouse Gas Initiative Program Overview, http://rggi.org/design/overview. Accessed Mar 2017.

61. RGGI 2012 Program Review: Summary of Recommendations to Accompany Model Rule Amendments, http://rggi.org/docs/ProgramReview/_FinalProgramReviewMaterials/Recommendations_Summary.pdf. Accessed Mar 2017.

62. Fact Sheet: Investment of RGGI Proceeds Through 2013, http://rggi.org/docs/ProceedsReport/Proceeds-Through-2013-FactSheet.pdf. Accessed Mar 2017.

63. P. J. Hibbard, S. F. Tierney, A. M. Okie and P. G. Darling, The Economic Impacts of the Regional Greenhouse Gas Initiative on Ten Northeast and Mid-Atlantic States, *The Analysis Group* (2011), http://www.analysisgroup.com/uploadedfiles/content/insights/publishing/economic_impact_rggi_report.pdf. Accessed Mar 2017.

64. Western Climate Initiative, Inc., http://www.wci-inc.org/. Accessed Mar 2017.

65. M. Purdon, D. Houle and E. Lachapelle, The Political Economy of California and Québec's Cap-and-Trade Systems, *Sustainable Prosperity* (2014), http://www.sustainableprosperity.ca/sites/default/files/publications/files/QuebecCalifornia%20FINAL.pdf. Accessed Mar 2017.

66. Summary Results Report, California Cap-and-Trade and Québec Cap-and-Trade System, February 2016 Joint Auction #6, https://www.arb.ca.gov/cc/capandtrade/auction/feb-2016/summary_results_report.pdf. Accessed Mar 2017.

67. Auction Proceeds Allocated to the Green Fund, http://www.mddelcc.gouv.qc.ca/changements/carbone/revenus-en.htm. Accessed Mar 2017.

68. J. Drance, J. Kroft and L. Sinclair, Determining the Effective Price of Carbon, *Canadian Energy Law* (2016), http://www.canadianenergylaw.com/tags/cap-and-trade/. Accessed Mar 2017.

69. Under2MOU, http://under2mou.org/. Accessed Mar 2017.

70. E. Hardball, "Gov. Cuomo aims to link U.S. Northeast's carbon market with Calif.'s", E&E News Climatewire, https://www.eenews.net/stories/1060026140. Accessed Mar 2017.

71. Cleaner Power Plants, https://www.epa.gov/mats/cleaner-power-plants. Accessed Mar 2017.

72. Clean Air Act Hazardous Air Pollutants, 42 U. S. C. §7412(b), https://www.law.cornell.edu/uscode/text/42/7412. Accessed Mar 2017.

73. 65 FR 79826, 79830 (2000), https://www.gpo.gov/fdsys/pkg/FR-2000-12-20/pdf/FR-2000-12-20.pdf. Accessed Mar 2017.

74. Court Opinion, State of New Jersey v. EPA, No. 05-1097, U.S. Court of Appeals for the District of Columbia Circuit, 2008, https://www.cadc. uscourts.gov/internet/opinions.nsf/ 68822E72677ACBCD8525744000470736/$file/05-1097a.pdf. Accessed Mar 2017.

75. 77 FR 9363 (2012), https://www3.epa.gov/airtoxics/utility/fr16fe12.pdf. Accessed Mar 2017.; Cleaner Power Plants, https://www.epa.gov/mats/ cleaner-power-plants. Accessed Mar 2017.

76. 77 FR 9326 (2012), https://www.gpo.gov/fdsys/pkg/FR-2012-02-16/pdf/ 2012-806.pdf. Accessed Mar 2017.

77. 77 FR 9306 (2012), https://www.gpo.gov/fdsys/pkg/FR-2012-02-16/pdf/ 2012-806.pdf. Accessed Mar 2017.

78. Michigan v. Environmental Protection Agency, No. 14-46, Supreme Court of the United States, 2014, https://www.supremecourt.gov/opinions/ 14pdf/14-46_10n2.pdf. Accessed Mar 2017.

79. Supplemental Finding That It Is Appropriate and Necessary To Regulate Hazardous Air Pollutants From Coal- and Oil-Fired Electric Utility Steam Generating Units, https://www.federalregister.gov/documents/2016/04/ 25/2016-09429/supplemental-finding-that-it-is-appropriate-and-necessary-to-regulate-hazardous-air-pollutants-from. Accessed Mar 2017.

80. http://blogs.edf.org/climate411/files/2016/11/Murray-Energy-v-EPA-Petitioners-combined-opening-brief-11-18-16.pdf?_ga=1.106881344. 238617807.1491062437. Accessed Apr 2017.

81. What is Acid Rain?, https://www.epa.gov/acidrain/what-acid-rain. Accessed Mar 2017.

82. Title IV – Acid Deposition Control, https://www.epa.gov/sites/ production/files/2015-06/documents/title_iv_-acid_deposition_control. pdf. Accessed Mar 2017.; https://www.epa.gov/airmarkets/acid-rain-program-laws-and-regulations.

83. Acid Rain Program Laws and Regulations, https://www.epa.gov/ airmarkets/acid-rain-program. Accessed Mar 2017.

84. The other main precursor, NOx, was handled with more traditional specific Nox emission limts at the source.

85. The Invisible Green Hand, *The Economist*, 4 Jul 2002, http://www. economist.com/node/1200205. Accessed Mar 2017.

86. R. Schmalensee and R. N. Stavins, The SO2 Allowance Trading System: The Ironic History of a Grand Policy Experiment, MIT Center for Energy and Policy Research, http://web.mit.edu/ceepr/www/publications/ workingpapers/2012-012.pdf. Accessed Mar 2017.

Liquid Fuels and Chemical Feedstocks

COLIN E. SNAPE

ABSTRACT

Coal can react directly with hydrogen to cleave C–C bonds and remove heteroatoms (O, N and S), which is referred to as "direct liquefaction". Such processes are catalyzed and involve more than one stage and generate distillates in high yield. Alternatively, coal can be gasified with steam into a mixture of carbon monoxide and hydrogen, known as synthesis gas or "syngas", which then can be converted with suitable catalysts into a wide variety of fuels and chemical feedstocks by what is commonly referred to as Fischer–Tropsch or FT synthesis. This route is known as "indirect liquefaction", which is mature in the sense that commercial plants are currently operating using this process, notably the SASOL process in South Africa. This contrasts with direct liquefaction where, despite the intensive process development that has occurred with a number of demonstration plants being operated, high capital costs and current low oil prices have prevented commercial plants coming into operation using this process, with the exception of the Shenhua plant in China. Despite the uncertain outlook, a major development programme since the 1970s, triggered by rapid increases in oil prices, has led to the severity of the process conditions being reduced considerably and a much greater understanding of the underlying chemistry has been gained, which provides the main focus for this overview. The impact of coal structure, solvent composition and catalysis on the conversion of coal to liquids is described, together with the chemical nature of the heavy intermediates and final distillate products.

Issues in Environmental Science and Technology No. 45
Coal in the 21st Century: Energy Needs, Chemicals and Environmental Controls
Edited by R.E. Hester and R.M. Harrison
© The Royal Society of Chemistry 2018
Published by the Royal Society of Chemistry, www.rsc.org

1 Introduction

1.1 Organic Structure of Coal

The organic structure of coal consists primarily of carbon, hydrogen and oxygen with smaller amounts of oxygen and sulfur being present and is the result of the combined action of temperature and pressure over geological timescales. Higher plant debris in the form of peat are transformed into brown coal and lignite, then sub-bituminous and bituminous coals and then anthracite.[1] The maturation process is referred to as "coalification" and the maturity of coal is usually measured by rank which can be depicted on a Van Krevelen plot of atomic H/C vs O/C ratio.[2,3] Figure 1 shows the transformation of coal from lignite through the bituminous rank range to anthracite.

Coal is largely macromolecular (kerogen) with only relatively small proportions of the organic matter being extracted in common organic solvents and, by virtue of being derived from woody biomass, is aromatic in character and contains oxygen functional groups. As rank or maturity increases, the proportion of aromatic carbon increases and oxygen functional groups decrease. Coal is heterogeneous in nature, with distinct entities known as "macerals" being discernible by optical microscopy using polished blocks under reflected light. The major maceral group for humic coals is "vitrinite", derived from cell walls in woody tissue. "Liptinite" macerals include waxes and resins from spores and cuticles; resins from plant tissues are generally less aromatic, reflecting less light than vitrinite. "Inertinite" macerals are the charred remains from fires and their concentrations vary considerably.

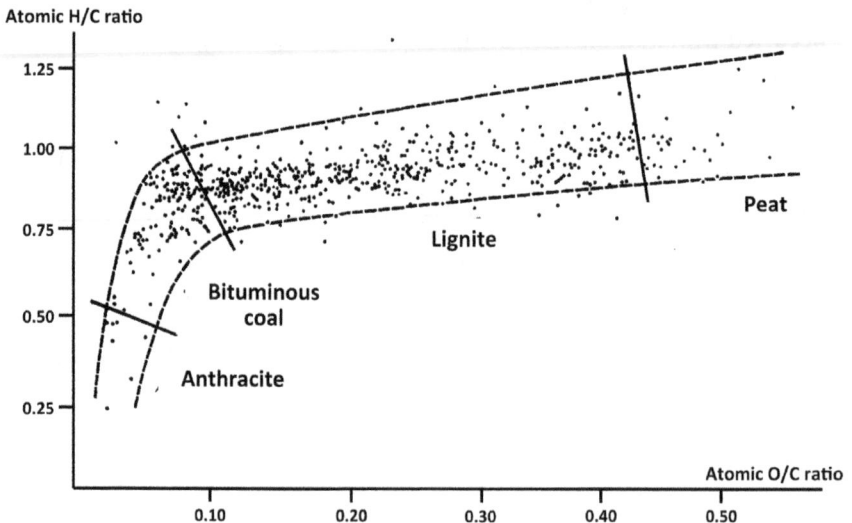

Figure 1 Van Krevelen diagram showing coal maturation as a function of rank. (Reproduced with permission from Durand and Paratte, *Geological Society, London, Special Publications*, 1983, **12**, 255–265).[3]

The heterogeneity of coals makes it impossible to define a unique structure for coal but "average structures" that match the overall structural
characteristics, including the fraction of aromatic carbon determined by
techniques such as solid state ^{13}C nuclear magnetic resonance (NMR) and
the distribution of oxygen functionalities determined by titration and
spectroscopic methods, can be deduced.[4,5] Structures such as that shown in
Figure 2 are a convenient way of conveying the highly aromatic nature of
coal, containing ring structures of varying size with aliphatic substituents
being a mixture of short alkyl chains, dominated by methyl and cyclic or
naphthenic groups. In highly volatile bituminous coals, typically 75–80% of
the carbon is aromatic. Phenols are the major oxygen form in bituminous
coals, where nitrogen and sulfur occur in aromatic structures, aza-bases plus
aromatic secondary amines and thiophenes, respectively. The heteroatomic
polar functionalities mean that H-bonding occurs to a significant degree.

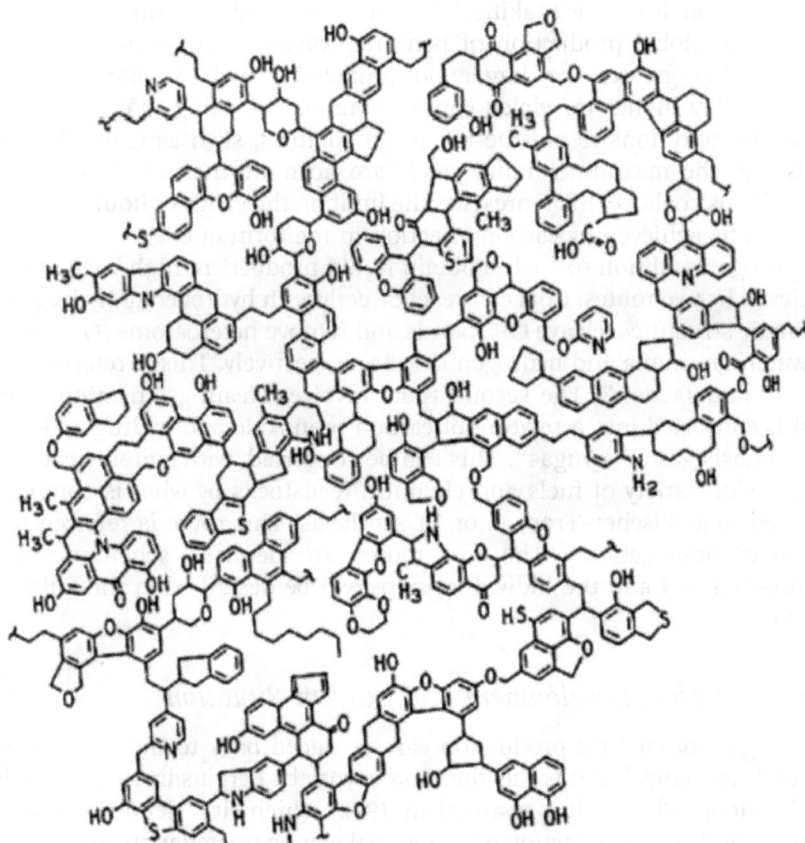

Figure 2 Model structure for bituminous coal.
(Reprinted from Fuel, 63(9), J. H. Shinn, From coal to single-stage and two-
stage products: A reactive model of coal structure, 1187–1196, Copyright
1984, with permission from Elsevier).[5]

Detailed information on the organic structure of coal can be found in Van Krevelen and other authoritative texts.[1,6-8]

1.2   Liquefaction Routes

"Liquefaction" refers to coal being converted to liquid products, transport fuels and chemical feedstocks in high yield and this is often referred to as "Coal to Liquids" or CTL. The aromatic nature of coals means that they have low atomic H/C ratios, which are typically 0.75-0.85 for bituminous coals with high volatile matter contents (see Figure 1). To achieve high conversions to liquid products, bearing in mind that the hydrogen contents of gasoline and diesel are 11-14% w/w, means that either hydrogen needs to be added or carbon needs to be rejected. Coal tar, which is a by-product of coke making, historically has played an important role in the origins of the chemical industry but yields are below *ca.* 10% w/w from traditional carbonization for coke making.[9] Nevertheless, coal tar still accounts for 10-15% of global production of benzene, toluene and xylenes (BTX) and considerably greater for larger polyaromatics, such as naphthalene.[10] Considerably higher tar yields can be obtained by using more favourable pyrolysis conditions to aid the release of volatiles, such as using fluidized beds, but the maximum liquid yields are no more than *ca.* 25% w/w for bituminous coals.[11] This represents the limit on the yields of liquid products that can be achieved by carbon rejection in the form of coke.

Hydrogen addition to coal to obtain liquid products in high yields can be achieved by two routes. Coal can react directly with hydrogen to hydrogenate aromatic structures, cleave C–C bonds and remove heteroatoms (O, N and S) as water, ammonia and hydrogen sulfide, respectively. This is referred to as "direct liquefaction". The second route involves steam gasification, where coal is converted into a mixture of carbon monoxide and hydrogen, known as synthesis gas or "syngas". This can be converted, with suitable catalysts, into a wide variety of fuels and chemical feedstocks by what is commonly referred to as Fischer–Tropsch or FT synthesis. This route is referred to as "indirect liquefaction". The two routes are depicted schematically in Figures 3 and 4 and the individual steps will be described in the following sections.

1.3   Historical Development and Current Situation

The origins of coal tar production can be traced back to the 1840s in the UK and Germany, but it is was the classic work by Bergius in 1913 (resulting in the Nobel Prize being awarded in 1932) which first demonstrated the concept of direct liquefaction whereby coal can be treated with solvents and hydrogen to generate liquid products in high yields.[12-15] This led to coal hydrogenation plants being operated during the 1930s in the Germany and the UK. Fischer–Tropsch (FT) synthesis was developed in Germany during the 1920s, resulting in gasoline being produced before and during World

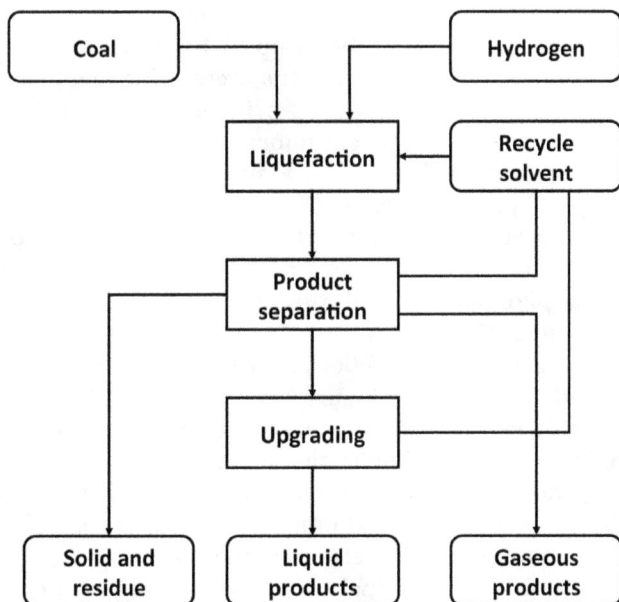

Figure 3 Simple representation of the steps involved in direct coal liquefaction.

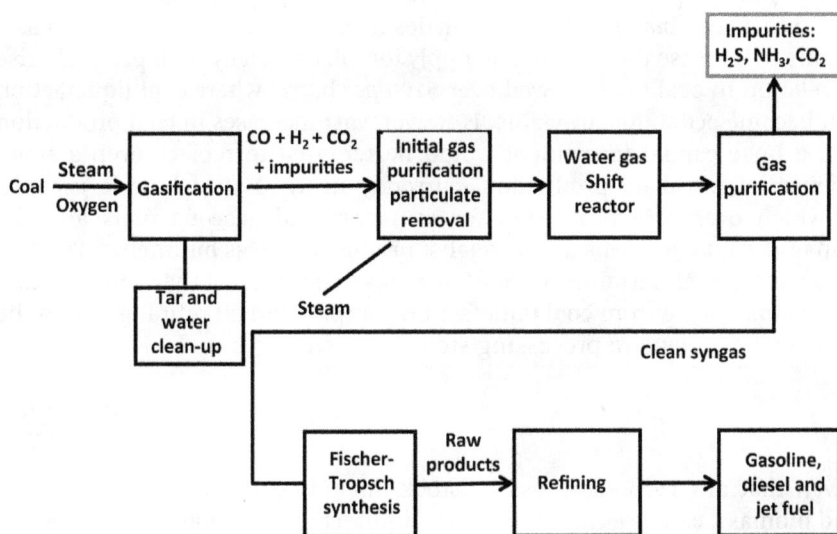

Figure 4 Simple representation of the steps involved in indirect coal liquefaction.

War II where production capacities reached 4 Mt *per annum* for direct and 0.6 Mt for indirect liquefaction,[14] which is clearly quite small compared to current global oil consumption (*ca.* 4 GT *per annum*). Interest in coal liquefaction waned during the late 1940s and 1950s due to the availability of cheap crude oil. However, in South Africa there was a need for a local supply of liquid fuels and this resulted in the development of the SASOL indirect liquefaction process. By the mid-1980s up to 10 Mt *per annum*, or 60% of South Africa's transport fuel requirements, were being met by these plants.

In the early 1970s the dramatic rises in oil prices resulted in major international efforts to develop coal liquefaction. Given that indirect coal liquefaction was already operating on a commercial scale,[12] the focus was on direct liquefaction processes using milder process conditions and optimizing the process steps compared to the early processes in the 1930s.[13-15] Common features in all the processes developed, a number of which were demonstrated at pilot-plant scale during the 1980s, are the dissolution of a high proportion of coal in a suitable solvent at elevated temperature, often assisted by hydrogen pressure, followed by hydrocracking of the dissolved coal with H_2 and a catalyst (see Figure 3).

Apart from the continued commercial operation of the SASOL process in South Africa, interest in coal liquefaction waned during the 1990s and 2000s with falling oil prices and proven reserves increasing, with vast discoveries in areas including Western Africa and Brazil. However, for countries with large coal and few oil reserves, coal liquefaction can clearly improve the security of supply for oil. China is the principal nation in this category and it has a large CTL development programme due to the increasing demand for transport fuels not likely to be met by indigenous oil production. A direct liquefaction demonstration plant is operated by Shenhua group, producing over 1 Mt of liquid fuels *per annum*.[15] Other countries may also see coal liquefaction as a means to increase the security of supply for oil, especially if oil prices do rise in relation to coal to levels well over $100 *per* barrel where coal liquefaction can become economically viable. However, vast increases in coal production and a huge capital investment would be required to replace even a small fraction of world oil production, currently nearly 100 M barrels *per* day, of which over 80% is utilized for transport and gaseous fuels and the remainder for chemicals and specialist products such as bitumen.[16] Further, regarding global warming, account now has to be taken of the increased CO_2 emissions arising from coal liquefaction compared to oil refining due to the more energy-intensive processing steps involved.[12,15]

1.4   Scope

Given that any carbonaceous feedstock, including natural gas, petroleum and biomass, can be gasified to syngas, indirect liquefaction is not unique to coal. Indeed, the use of natural gas and biomass are commonly referred to as "Biomass to Liquids" (BTL) and "Gas to Liquids" (GTL). Further, the severity of the gasification process means that the structure of the coal used only has a relatively small impact compared to direct liquefaction where the yields and compositions of the primary or intermediate products are heavily dependent on the rank and type of coal used. Therefore, after an overview of indirect coal liquefaction, the focus of the chapter is on direct liquefaction where the chemistry, catalysis and process steps are described.

2 Coal Gasification and Fischer–Tropsch Synthesis (Indirect Liquefaction)

2.1 Gasification

Synthesis gas or "syngas" is a mixture of carbon monoxide and hydrogen with carbon dioxide and, in principle, it can be produced from any carbonaceous material by steam gasification, which is highly endothermic – see eqn (1). If direct heating is used, then air or oxygen is introduced with steam to provide the process heat where eqn (2) and (3) indicate that roughly 30% carbon gasified needs to be combusted. Further, CO_2 formed can react with char to generate CO, which is also an endothermic reaction (3).

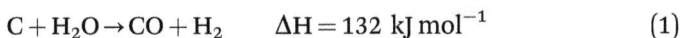

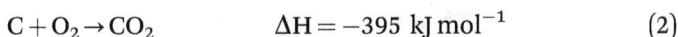

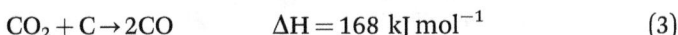

$$C + H_2O \rightarrow CO + H_2 \qquad \Delta H = 132 \text{ kJ mol}^{-1} \qquad (1)$$

$$C + O_2 \rightarrow CO_2 \qquad \Delta H = -395 \text{ kJ mol}^{-1} \qquad (2)$$

$$CO_2 + C \rightarrow 2CO \qquad \Delta H = 168 \text{ kJ mol}^{-1} \qquad (3)$$

Clearly, if air is used, a gas with low calorific value is obtained since nitrogen is the major constituent in the resultant gas. Therefore, oxygen has to be used to produce syngas. In contrast to coal and other solid reactants, indirect heating can be used for natural gas and this facilitates the use of catalysts, such as Ni, to reduce the reaction temperature.

Fixed and entrained flow reactors have been the most widely used for large-scale syngas production from coal.[17,18] Fixed-bed gasifiers were the first type used, by Lurgi in Germany, with a pressurised reactor developed in 1931.[18] Coal in lump form moves slowly downward under gravity and is gasified and partly combusted by counter flows of steam and oxygen, with tar being obtained as a by-product that is a source of chemicals (see Figure 5). An advanced form of the original reactor, the British Gas-Lurgi (BGL) slagging gasifier, was developed during the 1950s and 1960s, where ash is separated a molten slag.[17,18] Compared to early fixed-bed gasifiers, the lower steam and oxygen requirements give better quality syngas, containing less CO_2 and methane, and such gasifiers are currently used to supply syngas for FT synthesis in South Africa. For coals, entrained flow gasifiers with co-current flow have the advantage that any coal can be handled and tar production is avoided. Coal-water slurries are employed and the reactors operate over the pressure range 10–70 bar.[17,18] For power generation through integrated gasification combined cycle gasification (IGCC), they have been widely used in demonstration and commercial plants.

2.2 Water–Gas Shift Reaction

After initial gas cleaning to remove any solids, the water–gas shift (WGS) reaction, see eqn (4), can be used to increase the proportion of CO at the expense of hydrogen in the syngas.[8,9]

$$CO + H_2O \rightleftharpoons CO_2 + H_2 \qquad 41 \text{ kJ mol}^{-1} \qquad (4)$$

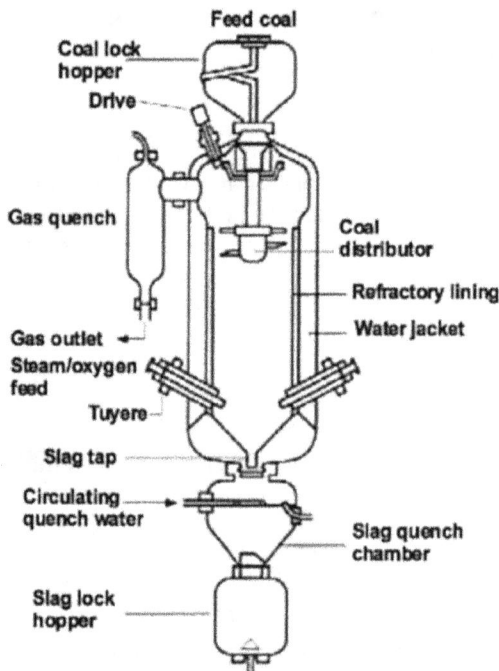

Figure 5 Schematic of the British Gas–Lurgi gasifier.
(Reprinted from Fuel, 84(17), A. J. Minchener, Coal gasification for advanced power generation, 2222–2235, Copyright 2005, with permission from Elsevier).[62]

There is flexibility in that the WGS reaction can be carried out before or after sulfur-containing gases are removed from sour and sweet gas, respectively. For sweet gas at temperatures above 350 °C iron oxide-based catalysts are used, as compared with sulfur-tolerant cobalt–molybdenum (Co–Mo) catalysts for sour gas.[19] At lower temperatures (*ca.* 200–250 °C), the equilibrium can be driven further with the copper–zinc oxide catalyst, which is similar to the catalysts needed to produce methanol from syngas that require extremely clean gas and are generally not used in coal gasification processes where sulfur-tolerant catalysts are required.[13,19]

Given that the WGS reaction is carried out under pressure, the CO_2 can be absorbed in a solvent which is alcohol-based and the pressure is then reduced to release the CO_2 in a cycle known as pressure-swing adsorption (PSA). If CO reacts completely to yield a mixture of CO_2 and hydrogen, PSA generates hydrogen with high purity that has a number of applications. It can be used instead of natural gas for power generation in combined cycle gas turbines that have high efficiency. Hydrogen at pressure is used in a number of refinery processes, including hydrocracking and hydro-desulfurization (HDS) to break heavy fractions down into distillate fuels and to remove organic sulfur, respectively, as will be described for direct liquefaction. Hydrogen can power fuel cells and can be used for heating instead of natural gas.[20] For all these applications, which collectively form part of the "hydrogen economy", if the separated CO_2 is captured and stored (see Carbon Capture and Storage in Chapter 7 of this book), then the hydrogen is considered as being "renewable" and this offers a major pathway to decarbonizing power generation, heat and transport.

2.3 Underground Coal Gasification (UCG)

Coal gasification can also be carried out underground. UCG provides a potential means for the economic recovery of energy from relatively deep coal deposits that are uneconomic to mine. UCG involves introducing steam and either air or oxygen into coal seams and the reactions involved broadly parallel those in normal gasification processes.[20,21] The first UCG trials date back to the early 20th century but despite many subsequent trials, there are few commercial operations; one of the few is in Angren, Uzbekistan, that has been operating for over 40 years and the gas provides operation fuel to a coal-fired power station.[20,21] Clearly, to produce syngas, oxygen needs to be used but, compared to normal gasification, the gas produced contains more methane and CO_2.

A typical UCG operation is depicted in Figure 6. Here, as for the shale gas exploration, the development of directional drilling has been critical, allowing large sections of coal to be gasified using just two boreholes. Directional drilling allows the point of injection being progressively

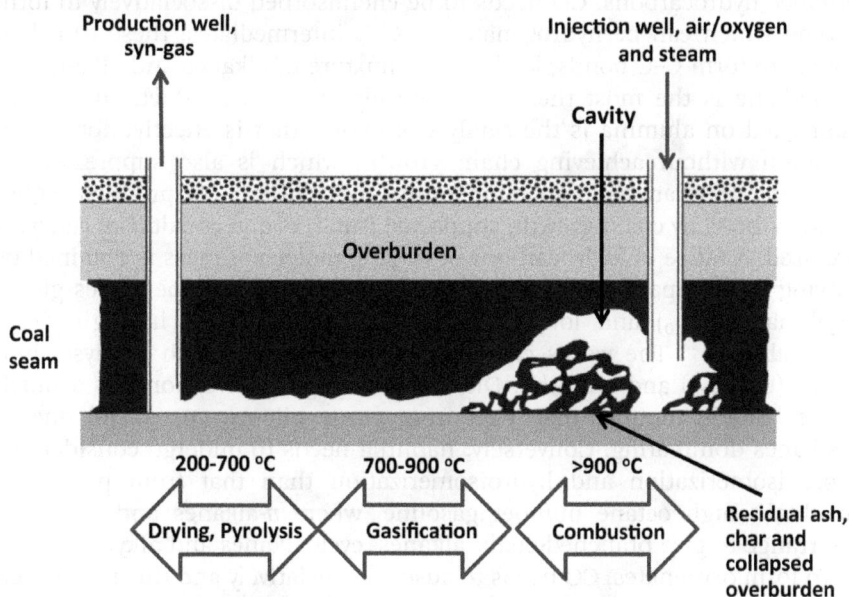

Figure 6 Schematic representation of underground coal gasification (UCG).

retracted away from the production well, enabling a relatively large area of coal to be extracted by this means. Once the amount of gas produced declines, a new reaction zone is established by moving the surface injection and production points. The formation of cavities can result in subsidence and this also provides a route for tar escaping to cause groundwater contamination and, consequently, relatively deep (typically over 500 m) operations are favoured on environmental grounds. There is also the advantage that the increasing pressure with depth enables the combustion and gasification reactions to proceed faster.

Clearly, to produce syngas, oxygen needs to be used, but, compared to normal gasification, the gas produced contains more methane (arising from the devolatization of the coal before gasification commences) and CO_2 due to generally a higher proportion of carbon being combusted than with the controlled conditions in normal gasification. However, the methane can be subsequently reformed with steam, the WGS reaction used and the CO_2 separated to produce syngas with the required molar fraction of hydrogen to CO.

2.4 Syngas Conversion

Syngas can be converted to a wide range of hydrocarbons and oxygenates.[13,22,23] Overall, syngas conversion is thermodynamically favourable for most products below 300 °C, with the reaction of CO and hydrogen being exothermic. The products obtained depend upon the transition metal catalysts, where both CO and hydrogen are chemisorbed, and the conditions used. To produce hydrocarbons, CO needs to be chemisorbed dissociatively to form carbon which can be hydrogenated to CH_x intermediates; these can then couple to form C–C bonds, leading to a mixture of alkanes and alkenes.

Methane is the most thermodynamically favoured product. Nickel (Ni) supported on alumina is the catalyst of choice that is effective for hydrogenation without achieving chain growth, which is also suppressed by operating at relatively high temperature (*ca.* 400 °C). To produce higher hydrocarbons by chain growth, supported iron (Fe) and cobalt (Co) catalysts are used. A range of hydrocarbons of varying molecular mass is obtained by varying the temperature in FT synthesis, with higher temperatures giving naphtha (C_5–C_{10}) and lower temperatures giving waxes in high yields (see Table 1).[23] The waxes can then be hydrocracked with catalysts into diesel (C_{10}–C_{20}) and naphtha. Diesel from syngas conversion has a much higher cetane number than that from crude oil (*ca.* 70, *cf.* 50) due to *n*-alkanes dominating. Conversely, naphtha needs to undergo considerably more isomerization and hydroisomerization than that from petroleum to obtain high octane number gasoline, where *n*-alkanes and *n*-alkenes re-arrange to give branched-chain alkanes, cycloalkanes and alkylbenzenes.

To form oxygenates, CO needs to adsorb associatively and this is favoured at lower temperatures. After associated chemisorption of CO, the reaction pathway involves propagation and termination steps as for hydrocarbon

Table 1 Product selectivities from low and high temperature Fischer–Tropsch (FT) synthesis for reactors in the SASOL process. (Reprinted from *Catalysis Today*, 23(1), B. Jager and R. Espinoza, Advances in low temperature Fischer–Tropsch synthesis, 17–28, Copyright 1995, with kind permission of Elsevier).[24]

Product (% w/w)	Low temperature	High temperature
Methane	4	7
C_2–C_4 alkenes	4	24
C_2–C_4 alkanes	4	6
Gasoline	18	36
Middle distillates	19	12
Heavy cut and waxes	48	9
Water soluble oxygenates	3	6
Main oxygenate components for the HTFT reactor (% w/w)		
Non-acids		
Acetaldehyde	3	
Acetone	10	
Ethanol	55	
Methyl ethyl ketone	3	
Isopropanol	3	
n-Propanol	13	
Isobutanol	3	
n-Butanol	4	
Acids		
Acetic acid	70	
Propionic acid	16	
Butyric acid	9	
Valeric acid and higher	5	

formation. The use of base promoters, such as K with Fe and Co catalysts, stabilizes the bond between the metals and CO and yields oxygenates in parallel to hydrocarbons, with a typical distribution of oxygen compounds obtained from FT synthesis being listed in Table 1.[13,22,24] Methanol is the least thermodynamically favoured common product in syngas conversion and requires a low temperature (250 °C). The choice of a catalyst that will avoid dissociative chemisorption is limited to copper (Cu) as the active component in the catalyst, which also contains alumina and zinc oxide (ZnO). As well as being a chemical feedstock, methanol can be converted to gasoline using a zeolite catalyst, ZSM5, where dimethylether is the intermediate product.[25]

3 Staging Direct Liquefaction and Primary Conversion

3.1 Concept

The direct conversion of coal requires the addition of significant amounts of hydrogen and processes differ in how this is achieved through the use of solvents, high pressure hydrogen and catalysis (see Figure 3), since it is not realistically possible to convert coal into premium transport fuels directly in a single step, given the stringent requirements on sulfur contents and other characteristics of gasoline, jet fuel and diesel. As with FT synthesis, there is some commonality with oil refining in that final processing is required to produce these transport fuels. It has been stated often that the thermal efficiency of direct coal liquefaction (the proportion of the heating value of the coal in the final products) is greater for the indirect route (*ca.* 65–70%, *cf.* 50–65%) due to the milder process conditions.[10,13] However, caution is needed in making such comparisons because efficiencies for direct liquefaction are for partially and not fully refined products as is the case for the indirect route.[26]

Solvents play a key role in being a source of hydrogen and dissolving the initial heavy liquid products obtained, as well as facilitating heat transfer. They can be used alone or in combination with high-pressure hydrogen and suitable catalysts, the latter process configuration giving rise to distillable liquids in high yields. Most processes are intended to be self-sufficient for both solvent and hydrogen gas, meaning that the solvent has to be generated from the liquid products and the quality of the solvent obtained by recycling is an important consideration. Undissolved coal and minerals later are separated from the primary conversion products, either by distillation if the products are relatively light, as for processes employing a combination of solvent, high pressure hydrogen and a catalyst, and by hot filtration or using solvents to precipitate ash and unconverted coal for milder processes where the primary products are subjected to further hydroprocessing (see Figure 3).

A number of reviews provide detailed technical information on many coal liquefaction processes that have been developed since the resurgence of interest during the 1970s.[13–15] Two distinct examples of the different process configurations are depicted in Figure 7. The Liquid Solvent Extraction (LSE) process is unique in that it only relies upon a hydrogen donor to liquefy coal, with the primary conversion product being upgraded to distillable products *via* catalytic hydroprocessing in the second stage (see Figure 7A). The H-Coal process developed by Hydrocarbon Technologies Inc (HTI) (see Figure 7B) is an example of a single-stage process that uses recycled solvent, hydrogen pressure and a hydroprocessing catalyst. The HTI process involves the same configuration to liquefy coal but with a dispersed rather than a supported catalyst (see Section 3.4) and a second stage to hydrogenate the recycled solvent to improve its hydrogen-donor properties. This is also similar to that being used for the Shenhua demonstration plant in China and to the NEDOL process developed in Japan,[27] a process developed specifically for brown coal.

Conversions are measured in terms of solubility and the solvents used with increasing polarity; *n*-alkane (*n*-pentane, *n*-hexane or *n*-heptane, $nC_5–nC_7$), toluene, tetrahydrofuran and either pyridine or quinoline are used for this purpose. The latter have been widely used to measure primary conversions and, as more hydrogen is consumed, primary products decrease in molecular mass with lower heteroatom concentrations and this manifests

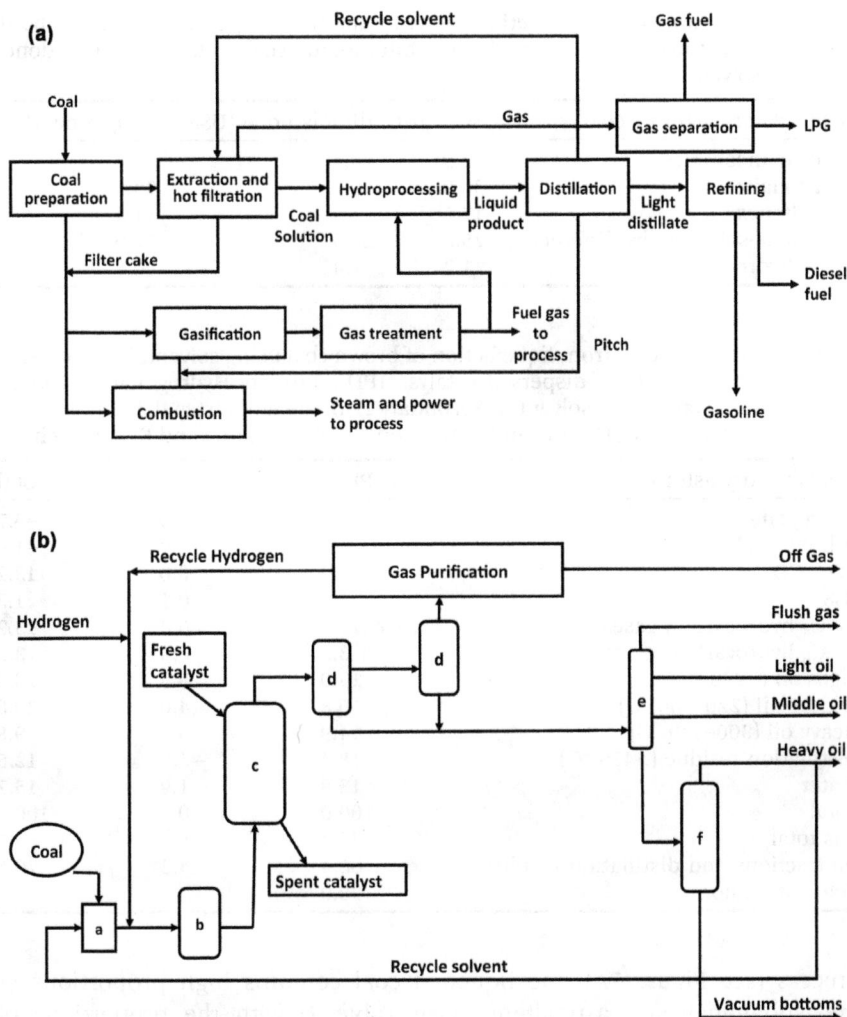

Figure 7 Process configurations for some direct coal liquefaction processes. (a) Liquid Solvent Extraction Process. (b) Schematic diagram of the H-Coal process: a = slurry mix tank; b = slurry heater; c = ebullated-bed reactor; d = separator; e = atmospheric distillation; and f = vacuum distillation. This is very similar to the NEDOL process.
(Data taken from ref. 13).

itself in a shift towards nC_5–nC_7 solubles (maltenes) and toluene-solubles/ toluene-insolubles (asphaltenes) at the expense of pyridine-soluble/toluene-insolubles (pre-asphaltenes). These are operational definitions and do not define specific chemical compositions. Further, a clear distinction needs to be made between primary conversions and oil or distillate yields, which are not necessarily related. When only solvents are used without high-pressure hydrogen in the primary stage, as in the Liquid Solvent Extraction (LSE)

Table 2　Comparison of liquefaction yields for an Arctic perhydrous coal, Svalbard, with those for a USA and a UK bituminous coal with a hydrogen donor solvent.

Yields (% dry ash free coal)	Svalbard	Illinois no. 6 (USA)	Longannet (UK)
Pyridine insolubles	12.9	29.4	41.4
Pre-asphaltenes	38.0	19.2	23.5
Asphaltenes	17.6	21.5	11.4
n-Heptane-solubles (by difference)	25.3	23.5	17.9
C_1–C_4 hydrocarbon gases	1.2	1.4	0.8

Table 3　Product yields from liquefaction of brown coal using solvent, high-pressure hydrogen and a dispersed catalyst. (PH = Primary Hydrogenation conversion stage; SH = solvent or Secondary Hydrogenation stage). (Adapted from I. Mochida, O. Okuma and S.-H. Yoon[17] and O. Okuma and K. Sakanishi[28]).

Yields (% dry ash free coal)	PH	SH	Total
Hydrogen	−4.7	−1.0	−5.7
Sulfur	−1.0	0.0	−1.0
$CO + CO_2$	13.2	0.0	13.2
H_2S	1.2	0.1	1.3
C_1–C_4 hydrocarbon gases	13.2	0.0	13.2
C_5–C_6 hydrocarbon gases	3.3	0.0	3.3
Light oil (<220 °C)	13.0	5.3	18.3
Middle oil (220–300 °C)	16.8	4.0	20.8
Heavy oil (300–420 °C)	14.9	−5.1	9.8
Distillation residue (>420 °C)	18.4	−5.9	12.5
Water	13.8	1.9	15.7
Total	100.0	0	100.0
Gas total	25.4		
Oil fractions and distillation residue	66.4	4.2	70.6
Coal conversion	98.0		

process (see Figure 7A), the liquefied coal contains high proportions of pre-asphaltenes and asphaltenes (see Table 2) with the proportions of maltenes or oils increasing when hydrogen pressure and catalysts are employed. In contrast, for processes employing solvent, high-pressure hydrogen and catalysts, the liquid products are predominately maltenes with high yields of distillate products, as demonstrated in Table 3 for brown coal where the distillate cuts account for over 40% w/w of the coal.[27,28]

3.2　Reaction Mechanisms

Coal dissolution is predominately a free radical process that can be represented simply by cleaving C–C and carbon–heteroatom bonds and the radicals being stabilized by hydrogen transfer, either from the liquid or gas phase.[29] More stable C–C bonds are cleaved by hydrogenolysis, where H transfer to an aromatic ring weakens the adjacent bonds. The breakdown of the molecular structure of coal results in intermediate products that are soluble in common organic solvents that are used to determine overall conversions. With increasing severity of the reaction conditions, these are further broken down into smaller molecules.[5]

3.3 Solvents

Solvents play a key role in mediating hydrogen transfer and for processes where high-pressure hydrogen is not employed in the first stage, such as the Liquid Solvent Extraction (LSE) process developed in the UK (see Figure 7A), the solvent is the only source of hydrogen and is thus referred to as a "hydrogen donor". Given that the solvent is recycled, with a boiling range typically 250–450 °C, it is important that the hydrotreating conditions in the second stage of such processes result in the right level of hydrogenation.

Partially hydrogenated aromatics, known as hydroaromatics (such as tetrahydronaphthalene (tetralin) and tetra/octahydrophenanthrenes), donate hydrogen readily, but fully hydrogenated multi-ring cycloalkanes or naphthenes (such as decahydronaphthalenes (decalin) and perhydrophenanthrenes), are poor hydrogen donors. The concentration of donatable hydrogen can be estimated by NMR, which has been used to compare the ^{1}H spectra of a good quality solvent with the residual solvent recovered after liquefaction, showing the virtually complete loss of aliphatic hydrogen from hydroaromatic species.[30] For good quality solvents, such as those used in the LSE process, containing *ca.* 2.0% w/w donatable hydrogen, if a typical solvent-to-coal mass ratio of 2 : 1 is used for liquefaction, 5.0% w/w of hydrogen on a coal mass basis will be available. Over many cycles, the performance of hydrogen donor solvents can deteriorate through re-arrangement reactions, such as the isomerization of tetralin to methylindan, which is not a hydrogen donor.[31]

For processes using a combination of hydrogen pressure and catalysts for liquefying coal (*e.g.* processes as shown in Figure 7), the concentrations of hydroaromatics in the solvents do not have to be as high as for the LSE process not employing hydrogen pressure, since polyaromatics are hydrogenated *in situ* to generate hydroaromatics. However, experience has shown a separate solvent hydrogenation step is beneficial in such processes (HTI and NEDOL) because of the competing demands to generate distillate products by hydrocracking (C–C bond cleavage) from the liquefied coal and solvent hydrogenation to provide hydroaromatics in the recycle solvent, with lower temperatures required for the latter.[13,15,26]

3.4 Catalysis

Two distinct types of catalysts, dispersed and supported, are used to aid the breakdown of the initial high molecular mass material from liquefying coal into lighter fractions in processes where high-pressure hydrogen is employed. The former are used to promote the primary hydrogenation of coal while the latter are most effective for upgrading coal-derived liquids. Disposable catalysts used are low cost and economically it is not worth considering their recovery from the unconverted coal and mineral matter remaining after liquefaction. Supported metal catalysts are used in ebullated bed reactors with continuous replacement, such as in the H-Coal process (see Figure 7B).[32,33] Given that the supported catalysts based on Ni–Co and molybdenum (Mo) are the same as used in petroleum refining and for upgrading coal liquids, these are covered in Section 4.

The activity of iron catalysts is generally low, necessitating the use of relatively high concentrations (several % w/w) of iron. Pyrite (FeS_2) and other iron compounds have attracted widespread interest due to their low cost but other metals, including tin and molybdenum, can be active in low concentrations.[13] For example, in the NEDOL process, pulverized natural pyrite is used at a concentration of 2–3 wt% of the coal mass.[34] The activity of iron-based catalysts depends upon their composition and dispersion in coal. A number of approaches have been employed to generate active dispersed iron catalysts, including mixing small particles into slurries of coal and solvent,[35] impregnating coal with water-soluble iron species, and using oil-soluble naphthenates and carbonates.[36] Table 4 indicates that coal impregnation with iron oxide precursors result in higher conversions than simple mixing of powdered iron compounds with coal arising from the resultant smaller 10–40 nanometer-sized iron sulfide crystallites formed.[37]

As stated above, iron has received much attention due to its low cost and availability, but other metals including Mo and halogen-promoted tin (Sn) can also be effective in relatively low concentrations, with sulfided Mo catalysts being described further in Section 4.

3.5   Impact of Coal Structure on Conversion and Product Composition

No one property can adequately predict conversions for all coals, but the characteristics of the coal play a more important role when dissolution is carried out under relatively mild conditions (*i.e.* at short residence times), with differences in conversion being generally less evident with increasing process severity.[38] However, as expected, it has been established that under a

Table 4   Impact of preparation on conversions for three bituminous coals with a non-hydrogen donor solvent comprising alkylnaphthalenes at 400 °C with a solvent-to-coal mass ratio of 2:1 and a hydrogen pressure of 69 bar. The hydrogen donor solvent was hydrogenated anthracene oil with a solvent-to-coal mass ratio of 2.5:1 at 420 °C for 1 hour in batch autoclave tests. (Adapted with permission from ref. 37. Copyright (1994) American Chemical Society).

Coal form	Coal conversion (% w/w dry ash free basis)	
	THF-solubles	*n*-Heptane-solubles
No catalyst	54–58	25–30
Physically mixed FeOOH	64–67	34–40
Impregnated FeOOH	79–85	39–50

wide range of liquefaction conditions, vitrinite and liptinite group mac-erals,[39,40] often referred to as "reactive macerals", from sub-bituminous and bituminous coals are generally more reactive than inertinites that have lower atomic H/C ratios. Indeed, reactive maceral contents have been found to correlate with conversions for coals of similar rank from particular geo-logical regions. However, in non-carboniferous coals from the Southern Hemisphere, inertinite, particularly the maceral semi-fusinite with a low reflectance, can contribute significantly to overall conversions and much better correlations are achieved when semi-fusinite is counted as a reactive maceral.[40,41]

The effect of coal rank on primary conversion shows completely different trends for hydrogen donor and non-hydrogen donor solvents. Conversions remain high from lignites through to bituminous coals for donor solvents until relatively high rank is reached. In contrast, for non-donor solvents, including polycyclic and alkyl substituted aromatics, highest conversions are achieved for coals containing *ca.* 87–89% C (see Section 4.1).[43] This corresponds to the point where coals display highest fluidity on heating, which corresponds to maximum in pyridine-extractable material.[1] Coalification involves elimination of oxygen, giving rise to increasing amounts of lower molecular mass solvent-extractable material. Eventually aromatization and ring growth reverses this trend, resulting in a decrease in pyridine-extractables.

When yields of distillate products are considered rather than total conversions, reasonable correlations have been established with atomic H/C ratios.[38] Therefore, both in terms of maximizing liquid yields and minimizing hydrogen consumption, there has been considerable interest in hydrogen-rich or perhydrous coals. Such coals are either rich in liptinite or contain hydrogen-rich degraded vitrinite, arising from the input of bacterial matter.[44] Table 2 indicates that both overall conversions to pyridine and toluene-soluble material and oil (maltenes) yields are higher for a perhydrous coal (Svalbard) than for typical bituminous coals.

There has been considerable interest in the composition of the primary or initial liquid products both to understand process chemistry and as a means to understand the structure of coal.[45–47] Indeed, the average chemical structures of asphaltenes, based on the aromatic and aliphatic carbon and hydrogen distributions derived from ^{1}H and ^{13}C NMR data and other structural parameters comprise small aromatic rings with associated short alkyl chains and naphthenic rings; these are considered to be representative of the structure of vitrinite, the major maceral in bituminous coals.[46,47]

High resolution mass spectrometric techniques have developed to the extent that they can define the molecular mass range and assign formulae to all species of a particular molecular mass, which provides a full structural definition of coal liquefaction products. Electrospray ionization (ESI) Fourier transform ion cyclotron resonance mass spectrometry (ESI-FTICR-MS) enables literally thousands of species to be identified and,[48] as an example of such detailed compositional data, Figure 8 shows the relative abundances of families of heteroatomic species.[48]

Relative Abundance%

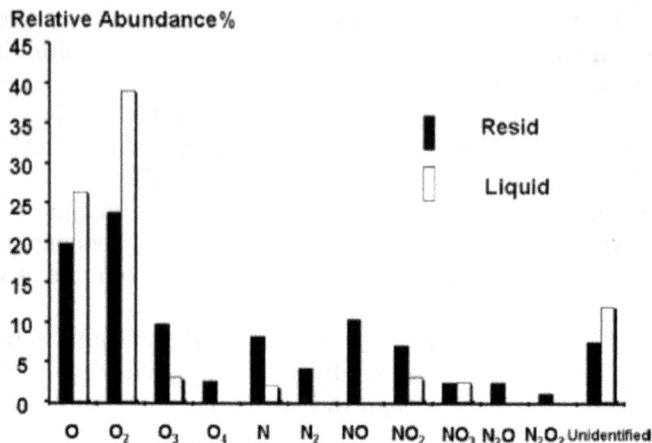

Figure 8 Relative abundance of different heteroatom-containing classes of compounds in a distillation residue (11 classes) and a distillable liquid fraction (6 classes). For simplicity, classes (shown on the x-axis) are abbreviated to represent only the heteroatom content. For example, naphthenic acids are abbreviated to O2 and pyrroles to N.
(Reprinted from Fuel, 84(14–15), Z. Wu, R. P. Rodgers and A. G. Marshall, ESI FT-ICR mass spectral analysis of coal liquefaction products, 1790–1797, Copyright 2005, with permission from Elsevier).[48]

4   Upgrading Intermediate Products to Transport Fuels and Chemicals

4.1   Hydroprocessing Heavy Coal Liquids

To upgrade the primary or initial products on liquefying coal, a catalyst needs to have activity for hydrocracking (cleaving C–C bonds), hydrogenation and heteroatom removal. Collectively, these reactions are known as hydroprocessing, with hydrodesulfurization (HDS), hydrodenitrogenation (HDN) and hydrodeoxygenation (HDO) referring specifically to the removal of sulfur, nitrogen and oxygen, respectively. Given that the liquefied coal has been subjected to temperatures close to 400 °C, the heteroatoms are all associated with aromatic structures, phenols and furans for oxygen, the most abundant heteroatom, thiophenes for sulfur, and aza-bases and aromatic secondary amines for nitrogen. In contrast to HDS and HDO, to remove nitrogen from compounds such as quinolone and carbazole as examples of an aza-base and aromatic secondary amine, an aromatic ring needs to be hydrogenated first so as to create aliphatic C–N bonds of sufficiently low strength.

Group VIII transition metals (Ni; palladium, Pd; and Pt) are effective hydrogenation catalysts and should be considered for HDS, HDN and HDO. However, transition metals generally react readily with H_2S to yield the corresponding metal sulfides with low catalytic activity. However, other transition metal sulfides are excellent HDS catalysts because sulfur compounds can be chemisorbed and H_2 can be dissociated. It turns out that binary metal sulfides, in particular Co–Mo, are optimal for HDS, having enthalpies of formation that are neither too high nor too low for reactions to proceed. In addition to Co–Mo, the binary sulfides used for hydro-processing include Ni–Mo and nickel–tungsten (Ni–W), usually supported on γ-alumina, are widely used for hydrotreating in petroleum refining.[49,50] Typically, such catalysts contain about 2% Co or Ni and 10% Mo or W (mole ratio of about 2 : 1) and are pre-sulfided before use with hydrogen sulfide. For such binary systems, mixed sulfide phases can form, together with the individual sulfides. Thus, sulfided Co–Mo supported catalysts are likely to contain Co_9S_8, MoS_2 and some mixed Co–Mo–S phase.[49] W provides greater hydrocracking activity than Mo.

Supported catalysts are used in the form of granules or extrudates in trickle bed (fixed)-bed reactors for coal liquids, where the liquid feed and hydrogen flow co-currently through the catalyst bed. For heavy coal liquids, carbon deposition is rapid typically during the first day on stream and is accompanied by a loss of surface area.[51] After this initial rapid decline, further deactivation occurs more gradually due to metal deposition being slower from contaminants, including iron and titanium from soluble organometallic species.[52] The carbon can be removed periodically by controlled combustion but metal deposition leads to irreversible loss of activity, particularly for HDS.

Distillate fuels obtained after catalytic upgrading have relatively high H/C ratios (>1.5) with low concentrations of heteroatom contents (O, N and S typically below 0.1%).[13–15] The conditions needed to generate good quality recycled solvent can place constraints on the level of hydrogenation that can be achieved in some processes, such as for the LSE process (see Figure 7A).

4.2 Refined Distillate Fractions

Final refining is required to reduce O, N and S to ppm levels to meet the requirements for transport fuels, generate chemicals in high yield and adjust the octane and cetane numbers of gasoline and diesel, respectively. Commercial HDS and HDN catalysts similar to those used for upgrading the primary liquefied products can be reduced to achieve extremely low heteroatom contents.[49,50] These catalysts also serve to hydrogenate aromatic compounds and so improve the quality of jet fuel and diesel cuts. Compared to petroleum and the FT synthesis, the distinguishing feature of naphtha, jet fuel and diesel is that cycloalkanes or naphthenes are present in higher proportions in distillates from direct liquefaction processes,[53] although *n*-alkane concentrations do increase with decreasing rank in going from bituminous coal to brown coal.

For high performance jet fuel, direct coal liquefaction has an advantage over petroleum in that pyrolytic decomposition above 300 °C is considerably less because cycloalkanes and hydroaromatics, such as decalin and tetralin, are much more thermally stable than are *n*-alkanes or alkylaromatics, even at

450 °C.[54] In gasoline, cylcoalkanes, mainly cyclohexanes, have higher octane numbers than *n*-alkanes but much lower than those of alkylbenzenes (>100) and certain branched-chain alkanes, where 2,2,4-trimethylpentane is the reference hydrocarbon, having an octane number of 100 in terms of anti-knock performance in the standard engine tests. Therefore, dehydrogenation of cylcoalkanes to alkylbenzenes can be used to increase octane numbers, which can be achieved with the platinum (Pt) catalysts used for normal reforming of petroleum naphtha.[53]

Hydrocarbon gas yields are considerable from direct liquefaction (see Tables 3 and 4) and these can be steam cracked to generate predominately alkenes. Similarly, naphtha can be converted predominately to benzene and alkenes by this route.[53]

5   Other Process Variants

The discussion has concentrated on the conversion of coal to distillate fuels using a combination of a hydrogen donor solvent, high-pressure hydrogen and catalysts. Here some alternative processing routes are described, where coal is treated with non-hydrogen donor solvents, reacted directly with hydrogen and co-processed with other feedstocks, including petroleum fractions and wastes.

5.1   Non-donor Solvents

If coal is treated with non-hydrogen donor solvents, a pitch-like material can be obtained in fairly high yield that is essentially an ash-free low-sulfur solid fuel. Research in the late 1960s and early 1970s demonstrated that the extraction of coal in anthracene oil from coal tar, rich in polycyclic aromatic hydrocarbons, generates a pitch-like material.[42] This can be considered for the same applications as coal-tar pitch; for example, carbon fibre production.

The Hyper-coal process was developed in Japan using 1-methylnaphthalene and light cycle oil as non-donor solvents which can give conversions of *ca.* 50% with the high-ash residue still suitable for combustion and the ash-free heavy product suitable for direct firing into gas turbines to realise higher efficiencies than for conventional pulverized fuel combustion.[55] The Solvent Refined Coal process using hydrogen pressure was initially developed to produce a clean boiler fuel before being modified for direct liquefaction.[13–15] Supercritical solvents, such as toluene, have been investigated for coal liquefaction but yields are lower (*ca.* 35% extraction yield) than with hydrogen donor solvents, albeit the liquefied coal is similar in composition, comprising relatively small-ring aromatic structures.[46]

5.2   Direct Hydrogenation

Of course, coal can be reacted directly with hydrogen, a process known as "hydropyrolysis" or "hydrogasification", depending upon the pressure used.

The need to convert coal directly to methane (or synthetic natural gas, SNG) arose from recognition of the benefits of natural gas over traditional coal gas and uncertainty over supplies. Clearly, syngas can be used but coal can be reacted directly with hydrogen at elevated temperature (*ca.* 800 °C) in entrained-flow reactors to generate predominately methane in high yield with benzene, toluene and xylenes by-products, together with *ca.* 50% char that can be gasified to generate the process hydrogen.[47]

If considerably lower temperatures and dispersed catalysts are used, liquid yields are nearly comparable to those obtained by direct liquefaction. Sulfided Mo has proved to be one of the most effective dispersed catalysts that can be used in relatively low concentrations, being generated from precursors including ammonium dioxydithiomolybdate. Conversions for sulfided Mo in fixed-bed hydrogenation for a bituminous coal have shown that concentrations as low as 0.2% w/w Mo are effective.[56] To translate these results into continuous operation, a fluidized-bed reactor needs to be used to provide sufficiently long residence times for reaction. This has been possible for oil shales where a pilot plant has been operated at the Institute of Gas Technology (IGT) in the USA with a maximum operating pressure.[57] However, an advantage of hydropyrolysis is that a second stage hydrotreatment can be carried out on the tar vapours at temperatures close to 400 °C to generate distillate products directly without the need for condensation.[58] This concept has been demonstrated for biomass in the IGT pilot plant.[59]

5.3   Integrated Processing with Other Feedstocks

As already described, direct liquefaction processes are designed to be self-sufficient regarding their demands for energy, solvents and hydrogen. However, integrating direct liquefaction with oil refining means that this does not have to be the case and improved overall efficiencies can potentially be achieved by using refinery fractions in solvents and mixing the liquefied coal with petroleum in existing processes including naphtha reforming, hydrotreating and fluid catalytic cracking. In fact, if heavy petroleum fractions are used as solvents with high hydrogen pressure and dispersed catalysts, solvent recycling can be avoided completely in a once-through mode of operation.[60] However, this would require heavy petroleum fractions to be available in considerably greater quantities than the coals being liquefied.

Regarding the integration of direct liquefaction with waste processing, the LSE process has the advantage of not requiring high-pressure hydrogen in the first stage to aid coal dissolution. This opens up the possibility of using the process to liquefy wastes, such as sewage sludge, wood waste and plastics close to the point of their generation. Further, waste solvents can be considered for use in the process, including engine oils, fats and greases. Thus, one scenario is to use a combination of imported good-quality solvent containing relatively high concentrations of hydroaromatics with varying additions of waste solvent components and plastics, which will thermally decompose into solvent close to the source of the wastes. The primary liquid fuels with high energy density can be used directly as a heavy-fuel substitute or transported to existing oil refineries.

6 Concluding Remarks and Future Perspectives

Indirect liquefaction is mature in the sense that commercial plants are operating, which contrasts with direct liquefaction where, despite the intensive process development that has occurred, with a number of demonstration plants being operated, high capital costs and current low oil prices have prevented commercial plants coming into operation, with the exception of the Shenhua plant in China. As already highlighted, coal liquefaction is only likely to be implemented where concerns exist over the security of oil supply, particularly for China and historically for South Africa. Given that proven oil reserves have increased significantly over the past 20 years, it is uncertain whether many other counties are going to require coal liquefaction in the near future to ensure energy security.

The international drive to reduce CO_2 emissions to limit the atmospheric concentration to *ca.* 450 ppm, corresponding to global warming of no more than 2 °C above pre-industrial temperatures, means that it is likely that a significant proportion of current oil reserves will not be recovered, as fossil fuels are replaced by renewable alternatives. Further, where fossil energy continues to be used, there is going to be a drive to eliminate carbon emissions from their production by applying carbon capture and storage (CCS) to refinery operations. This will disadvantage coal liquefaction due to the CO_2 emissions being considerably higher than those from oil refining. However, most of the additional CO_2 emissions arise from hydrogen production, where the CO_2 separated following the WGS reaction that is currently vented to the atmosphere can easily be captured, but this requires oil refineries and liquefaction plants to be situated relatively close to storage sites for the CO_2. Integration with power generation is likely to offer synergies with respect to CCS.

This raises the question as to how coal liquefaction technologies might be deployed in the future. Clearly, indirect liquefaction can be applied to biomass, but large gasification plants are needed for economies of scale. The process concepts developed for direct liquefaction are also applicable to biomass where, as mentioned, the use of thermolytic solvent extraction can generate much superior intermediate oils in terms of reduced oxygen content compared to those obtained by pyrolysis.[61] Such products could feed into existing oil or future biofuel refinery operations as a means for demonstrating aspects of the technology.

Clearly, a range of chemicals are being generated by FT synthesis, these including oxygenates that cannot be obtained directly from petroleum, adding to the phenols and aromatics provided from coal tar which, of course, can be a by-product of gasification. Although direct liquefaction processes are not necessarily going to be established specifically for chemicals, they could well be used to generate fuels with specific properties, notably jet fuel with far superior thermal stability than that from petroleum. Finally, direct liquefaction has been one of the drivers to improve our knowledge and understanding of coal structure, which has benefits in other applications, for example, coke making, to understand the transformations that occur when bituminous coals undergo softening.

References

1. D. W. Van Krevelen, *Coal-Typology-Chemistry-Physics-Constitution*, Elsevier. Amsterdam, 2nd edn, 1981.
2. D. W. Van Krevelen, *Org. Geochem.*, 1984, **6**, 1.
3. B. Durand and M. Paratte, *Geological Society, London, Special Publications*, 1983, **12**, 255.
4. J. P. Mathews and A. L. Chaffee, *Fuel*, 2012, **96**, 1.
5. J. H. Shinn, *Fuel*, 1984, **63**, 1187.
6. R. C. Neavel, in *Chemistry of Coal Utilisation, Second Supplementary Volume*, ed. M. A. Elliott, John Wiley, 1981, pp. 91–158.
7. J. G. Speight, *Chemistry and Technology of Coal*, CRC Press, 3rd edn, 2012.
8. N. Berkowitz, *The Chemistry of Coal*, Elsevier, 1985.
9. D. McNeil in Chemistry of Coal Utilisation, Second Supplementary Volume, ed. M. A. Elliott, John Wiley, 1981, pp. 1003–1084.
10. H. C. Schobert and C. Song, *Fuel*, 2002, **81**, 15–32.
11. R. J. Tyler, *Fuel*, 1980, **59**, 218.
12. R. H. Williams and E. D. Larson, *Energy Sustainable Dev.*, 2003, 7(4), 103.
13. T. Kaneko, F. Derbyshire, E. Makino, D. Gray and M. Tamura, Coal Liquefaction, in *Ullmann's Encyclopedia of Industrial Chemistry*, John Wiley & Sons, Inc., 2005.
14. D. B. Dadyburjor and Z. Liu, Coal Liquefaction, in *Kirk–Othmer Encyclopedia of Chemical Technology*, Wiley-Interscience, New Jersey, 2004.
15. S. Vasireddy, B. Morreale, A. Cugini, C. Song and J. J. Spivey, *Energy Environ. Sci.*, 2011, **4**, 311.
16. *IEA Oil market Report*, 2015 Statistical Supplement.
17. A. J. Mitchener, *Fuel*, 2005, **84**, 2222.
18. R. W. Breault, *Energies*, 2010, **3**, 216.
19. J. C. W. Kuo, in *The Science and Technology of Coal and Coal Utilization*, ed. B. R. Cooper and W. A. Ellingson, Plenum, New York, 1984, ch. 5.
20. E. Shafirovich and A. Varma, *Ind. Eng. Chem. Res.*, 2009, **48**, 7865.
21. D. Yang, N. Koukouzas, M. Green and Y. Sheng, *J. Inst. Energy*, 2016, **4**, 469–484.
22. *Fischer-Tropsch Synthesis, Catalysts, and Catalysis: Advances and Applications*, ed. B. H. Davis and M. L. Occelli, CRC Press, 2016.
23. P. Gerard, A. A. C. Van Der Laan and M. Beenackers, *Catal. Rev.: Sci. Eng.*, 1999, **41**, 255.
24. B. Jager and R. L. Espinoza, *Catal. Today*, 1995, **23**(1), 17.

25. S. L. Meisel, J. P. McCullough, C. H. Lechthaler and P. B. Weisz, *Chemtech*, 1976, **6**, 86.
26. M. Hook and K. Aleklett, *Int. J. Energy Res.*, 2010, **34**, 848.
27. I. Mochida, O. Okuma and S.-H. Yoon, *Chem. Rev.*, 2014, **114**, 1637.
28. O. Okuma and K. Sakanishi, in *Advances in the Science of Victorian Brown Coal*, ed. C. Z. Li, Elsevier, New York, 2004, ch. 8, p. 441.
29. D. F. McMillen, R. Malhotra, G. P. Hum and S. J. Chang, *Energy Fuels*, 1987, **1**, 193.
30. J. W. Clarke, T. D. Rantell and C. E. Snape, *Fuel*, 1982, **61**, 707.
31. J. J. de Vlieger, A. P. G. Kieboom and H. Van Bekkum, *Fuel*, 1984, **63**, 334.
32. S. W. Weller, *Energy Fuels*, 1994, **8**, 415.
33. A. V. Cugini, D. Krastman, R. G. Lett and V. D. Balsone, *Catal. Today*, 1994, **19**, 395.
34. K. Ikeda, K. Sakawaki, Y. Nogami, K. Inokuchi and K. Imada, *Fuel*, 2000, **79**, 373.
35. G. T. Hager, X. X. Bi, P. C. Eklund, E. N. Givens and F. J. Derbyshire, *Energy Fuels*, 1994, **8**, 88.
36. T. Suzuki, O. Yamada, K. Fujita, Y. Takegami and Y. Watanabe, *Ind. Eng. Chem. Prod. Res.*, 1985, **24**, 832.
37. A. V. Cugini, D. Krastman, D. V. Martello, E. F. Frommell, A. W. Wells and G. D. Holder, *Energy Fuels*, 1994, **8**, 83.
38. C. E. Snape, *Fuel Process. Technol.*, 1987, **15**, 257–279.
39. J. F. Cudmore, *Fuel Process. Technol.*, 1978, **1**, 227.
40. D. Gray, G. Barrass, J. Jezko and J. R. Kershaw in *Coal Liquefaction Fundamentals*, ed. D. D. Whitehurst, American Chemical Society Symposium Series 139, 1980, pp. 35–82.
41. S. Heng and M. Shibaoka, *Fuel*, 1983, **62**, 610.
42. J. W. Clarke, G. H. Kimber, T. D. Rantell and D. E. Shipley in *Coal Liquefaction Fundamentals*, ed. D. D. Whitehurst, American Chemical Society Symposium Series 139, 1980, pp. 111–129.
43. F. J. Derbyshire and D. D. Whitehurst, *Fuel*, 1981, **60**, 655.
44. C. Marshall, D. J. Large, W. Meredith, C. E. Snape, C. Uguna, B. F. Spiro, A. Orheim, M. Jochmann, I. Mokogwu, Y. Wang and B. Friis, Geochemistry and petrology of Palaeocene coals from Spitzbergen–Part 1: oil potential and depositional environment, *J. Coal Geol.*, 2015, **143**, 22–33.
45. D. D. Whitehurst, T. O. Mitchell and M. Farcasiu, *Coal Liquefaction — The Chemistry and Technology of Thermal Processes*, Academic Press, New York, 1980.
46. A. A. Herod, W. R. Ladner and C. E. Snape, *Philos. Trans. R. Soc. London*, 1981, **A300**, 3.
47. C. E. Snape, W. R. Ladner, L. Petrakis and B. C. Gates, *Fuel Process. Technol.*, 1984, **13**, 155.
48. Z. Wu, R. Rodgers and A. G. Marshall, *Fuel*, 2005, **84**, 1790.
49. H. Topsøe, B. S. Clausen and F. E. Massoth, *Hydrotreating Catalysis*, Catalysis-Science and Technology Series, vol. 11, Springer, 1996.

50. M. S. Rana, V. Samano, J. Ancheyta and J. A. I. Diaz, A review of recent advances on process technologies for upgrading of heavy oils and residua, *Fuel*, 2007, **86**, 1216.
51. G. J. Stiegel, R. E. Tischer and L. M. Polinski, *Ind. Eng. Chem. Prod. Res. Dev.*, 1983, **22**, 411.
52. D. S. Thakur and M. G. Thomas, *Appl. Catal.*, 1983, **6**, 283.
53. B. L. Crynes, Processing Coal Liquefaction Products, in *Chemistry of Coal Utilisation, Second Supplementary Volume*, ed. M. A. Elliott, John Wiley, 1981, pp. 1991–2070.
54. L. M. Balster, E. Corporad, M. J. DeWitt, J. T. Edwards, J. S. Ervin, J. L. Graham, S. Young Lee, S. Pale, D. K. Phelps, L. R. Rudnick, R. J. Santoro, H. H. Schobert, L. M. Shafer, R. C. Striebich, Z. J. West, G. R. Wilson, R. Woodwarde and S. Zabarnick, *Fuel Process. Technol.*, 2008, **89**, 364.
55. N. Okuyama, N. Komatsu, T. Shigehisa, T. Kaneko and S. Tsuruya, *Fuel Process. Technol.*, 2004, **85**, 947.
56. C. E. Snape, C. J. Lafferty, H. P. Stephens, R. G. Dosch and E. Klavetter, *Fuel*, 1991, **70**, 393–395.
57. H. P. Stephens, C. Bolton and C. E. Snape, *Fuel*, 1989, **68**, 161–167.
58. M. J. Roberts, D. M. Rue and F. S. Lau, *Fuel*, 1992, **71**, 1433.
59. T. L. Marker, L. G. Felix, M. B. Linck and M. J. Roberts, *Environ. Prog. Sustainable Energy*, 2012, **31**, 191.
60. J. S. Speight and S. E. Moschopedis, *Fuel Process. Technol.*, 1986, **13**, 215.
61. H. Deng, W. Meredith, C. N. Uguna and C. E. Snape, *J. Anal. Appl. Pyrolysis*, 2015, **113**, 340.
62. A. J. Minchener, *Fuel*, 2005, **84**(17), 2222–2235.

Carbon Capture and Storage and Carbon Capture, Utilisation and Storage

E. J. "BEN" ANTHONY

ABSTRACT

The last several years have seen a significant withdrawal of support for Carbon Capture and Storage (CCS) demonstrations worldwide, with major projects being cancelled both in the USA and UK, among other jurisdictions. However, they have also seen the first serious demonstration of CCS technology for power production in Canada. Notwithstanding this seesawing of developments, CCS remains essential technology if the world is to meet its COP21 commitments and avoid ambient CO_2 levels in the environment of 450 ppm and above. In this respect, it seems almost certain that renewable energy developments will be insufficient to remain below the 450 ppm ceiling. This chapter will cover the history and current state of the art in CCS, as well as explore some of the complementary technologies.

1 Introduction

1.1 A Brief History of Global Warming Science, Starting with Tyndall and Arrhenius

The idea that gases in the atmosphere influence the earth's surface temperature is a venerable one. The first serious scientific work was carried out in 1859 by John Tyndall, an Irish physicist, who provided an experimental demonstration of the ideas of De Saussure and Fourier that the polyatomic

Issues in Environmental Science and Technology No. 45
Coal in the 21st Century: Energy Needs, Chemicals and Environmental Controls
Edited by R.E. Hester and R.M. Harrison
© The Royal Society of Chemistry 2018
Published by the Royal Society of Chemistry, www.rsc.org

gases in the atmosphere could influence surface temperature,[1] and by 1863 he had clearly suggested that changes in those gases could affect the earth's surface temperature. It was not, however, until 1896, that Savante Arrhenius carried out hand calculations demonstrating that a doubling of the atmospheric CO_2 might increase the earth's surface temperature by 3 or more degrees Centigrade, depending on latitude.[2] Then in his popular science book, "Worlds in the Making" published in 1907, he makes a clear link between climate change and coal burning and notes that "any doubling of the percentage of the carbon dioxide in the air would raise the temperature of the earth's surface by 4 degrees".[3]

Since that time there have been various vicissitudes in ideas about global warming (including the belief that it might be beneficial), with new advocates of the theory that man-made global warming was potentially significant coming to the fore, such as the English scientist Guy Callendar, despite a background of wide-scale scepticism.[4] However, in the last several decades a consensus has developed that global warming is a significant issue for humanity, with large potentially negative consequences.[5] There has also been an increasing effort to quantify the cost of global warming *versus* mitigation (which ultimately can be viewed as simply living with the consequences) in studies such as the *Stern* Review.[6] Cost analyses suggest a $3:1$ saving in terms of the cost of preventing climate change as opposed to the cost of mitigation strategies, although more significant ratios have since been suggested. There is also a recognition that quantifying the costs of global warming is difficult and likely to be partial, tending to underestimate the probable consequences for at-risk communities.[7] While there remains a vigorous debate about whether limiting growth is the best option to reduce anthropogenic emissions,[8] there seems little doubt that something must be done if we are to avoid very costly consequences.

1.2   Early Developments in Carbon Capture and Storage (CCS) Technology

Carbon Capture and Storage (CCS) or Carbon Capture and Sequestration, as it is sometimes called, represents an attempt to allow the use of fossil fuels by drastically reducing overall CO_2 emissions (by 90% or more) by first capturing and then storing that CO_2 in aquifers and other underground storage sites, thus dramatically reducing the rate at which CO_2 is added to the atmosphere by anthropogenic activity.[9]

Paul Collier[10] has produced two simple equations that attempt to represent the various interactions between the use of natural resources and societal costs:

$$\text{Nature} + \text{Technology} + \text{Regulation} = \text{Prosperity} \tag{1}$$

$$\text{Nature} + \text{Regulation} - \text{Technology} = \text{Hunger} \tag{2}$$

Alternatively, in the view of the author, one could rewrite eqn (1) as:

$$\text{Nature} + \text{Engineering} + \text{Regulation} = \text{Prosperity} \tag{3}$$

and eqn (2) as:

$$\text{Nature} + \text{Technology (Engineering)} = \text{Prehistory to the Present} \tag{4}$$

Carbon Capture and Storage attempts to achieve the state described by eqn (1).

1.3 An Outline of the Various Routes for Carbon Capture and Storage (CCS)

The three basic routes for capturing CO_2 are pre-combustion (essentially gasification technology), post-combustion capture, and oxygen firing to produce a pure stream of CO_2 suitable for storage or possibly use. In addition to these possibilities there are also other technologies such as chemical looping combustion,[9] in which the oxygen is separated from the air by means of a metal or metal oxide and brought to the combustion zone, thus producing a stream of condensable steam and almost pure CO_2 which can then be stored underground, potentially for extended periods (10^4–10^6 years).[11] In general, the consensus view is that separating the CO_2 or providing a pure stream of CO_2 suitable for storage is the most expensive step in CCS, representing perhaps 70–80% of the cost. This has meant that most of the research and engineering effort has focused on the carbon capture step, which is the subject of this chapter. It should be noted that the continued inclusion of CCS as an international policy option can also be regarded as a tacit recognition that world coal demand has been growing and can be expected to continue to grow in the next 20 years or so.[9] Current projections of world coal consumption suggest increases from 2012 to 2040 at an average rate of 0.6% *p.a.* from 161 EJ (EJ: exajoule, equivalent to 10^{18} J) in 2012 to 178 EJ in 2020 and to 190 EJ in 2040.[12] It should also be noted that no attempt is made here to include the CO_2 utilisation option, other than for enhanced oil recovery, on the basis that use by itself cannot provide an adequate replacement for CCS.

1.4 International Developments in the Deployment of Carbon Capture and Storage (CCS)

In September 2016, the Massachusetts Institute of Technology listed 13 such projects, but even since this date there has been a steady closure or postponement of projects.[13,14] For some countries such as the UK, closure of coal-powered facilities may represent a partial option to minimise CO_2 emissions and meet national and international targets.[15] Even natural gas will require CCS to meet such targets, unless nuclear and other renewable technologies can come to the fore in a major way. However, as noted previously, there appears to be little evidence of a dramatic decline in the use of fossil fuels worldwide presently, at least until 2050.[16] Currently, CCS appears to be a critical part of meeting the 2 °C scenario.[17] The alternative is to fail to meet such goals, possibly dramatically, and there is increasing concern that the western world is failing to meet these goals in countries such as the UK and the USA.[18,19] However, there have been arguments made that if the science associated with the technology were presented in the correct manner, the future of CCS should be successful.[20] Nonetheless, it must be admitted at this time that there is simply a dearth of such projects, even where extensive disposal sites exist.[21]

2   Pre-combustion, Post-combustion and Oxy-fuel Technologies

2.1   An Examination of the Potential of Pre-combustion Routes

Pre-combustion implies that the carbon is removed prior to the combustion process itself, and normally this is done by means of gasification, which can be regarded as the conversion of a carbonaceous fuel (liquid or solid) into a gas with a usable heating value.[22] At the utility scale the most important embodiment of this technology is integrated gasification combined cycle (IGCC), which is now a well-established technology in many industries. IGCC uses high-pressure O_2 to drive the gasification step. Typically, any CO_2 produced in the gasification step is then removed at high pressures by means of an organic solvent and the resulting syngas, composed primarily of a mixture of CO and H_2, is then burned in a turbine. If the goal is to capture all of the CO_2 produced in the process, then an additional step, such as the use of a shift reaction to produce H_2, must be employed, or a conventional amine scrubber must be used at the backend of the process. Currently, the use of IGCC for the conversion of coal for power generation is severely limited due to problems associated with the cost, reliability of the technology and overall availability of such units. The use of large-scale shift reactors to produce H_2 in industrial quantities from coal gasification is also still in the development stage, although it is extremely well established for other types of gasification. In general, the use of gasification technology for the production of chemicals, fertilisers, and transportation fuels continues to develop rapidly. The Gasification and Syngas Technologies Council notes that there are 272 operating gasification plants worldwide, with 686 gasifiers, and that another 74 plants are under construction, with China as the largest user of this technology.[23] By contrast, there are perhaps not more than half a dozen large units producing electrical power from coal and oil, with another 10 or so such projects currently in various stages of development.[24]

The number of biomass gasifiers is also low. However, given that biomass currently provides more than 10% of the global energy supply and ranks among the top four energy sources for world final energy consumption,[25] it is reasonable to suppose that biomass gasification will at some time become a significant route for reducing the anthropogenic CO_2 burden. This is especially likely if biomass gasification is practised with carbon capture: the so-called biomass energy with CO_2 capture and storage (BECS) option. Biomass also comes with a number of net negatives for large-scale use, including its relatively low energy density, potentially high moisture content and undesirable physical parameters (including anisotropic physical structure), all of which will discourage the use of gasification technologies.[26]

Currently, the processing of biomass into pellets (which have a low moisture content, consistent physical properties and a heating value near that of lignite) remains one of the better options for use in large conventional boilers of the type which could be fitted with amine scrubbing, or for a potential gasification step. However, considerable energy is used in pellet preparation and transportation, and one study, for instance, estimated that up to 40% of the energy is consumed in producing and transporting wood pellets from Canada to Europe.[27]

In the longer term, the possibility of the use of a mild heating process (at 200–300 °C), known as "torrefaction", also remains a potential route for treating biomass for large-scale applications.[28] Like pelletisation, torrefaction produces a substantial improvement in the physical form of biomass, making it more coal-like and eliminating its moisture content. As a consequence, torrefaction is envisaged as a process which can be used to produce large quantities of biomass for use in conventional power stations as a substitute for coal. Worldwide, it is expected there will be continued pressure to use such materials in both CO_2 avoidance and BECS strategies in the foreseeable future, possibly together with gasification technology and also with the use of solid oxide fuel cells (SOFC) to achieve higher efficiencies. Should SOFC become a major technology for the production of electricity, then the gasification route would represent the best use of these treated biomass fuels.[29]

2.2 Post-combustion Options

Currently, the option of post-combustion removal of CO_2 remains the main contender for CCS applications. That is the use of amine scrubbing to remove the CO_2 at near-ambient temperatures. Often, this option is combined with enhanced oil recovery (EOR), which will be discussed in Section 5.3. Currently, the largest such plant is the 110 MWe (MWe: megawatt electric which refers to the electrical output capability of the plant) Boundary Dam project in Saskatchewan, which has been operational since 2014.[30]

Amine scrubbing depends on the chemisorption of CO_2 from flue gas using a chemical sorbent.[31] Although the most well-known of these materials is an aqueous solution of monoethanolamine (MEA), in practice there is extensive development of new amine materials, which are often called advanced amines, and new amines or amine blends are under active investigation.[32] Current research is directed to improving these materials and, in the period of 2001 to 2012, improvements in sorbents have reduced the estimated work of separating CO_2 from coal-fired flue gas from 450 to about 200 kWh tCO$_2$$^{-1}$.[9] Typically, in such a system, about 90% of the CO_2 produced is expected to be captured.[33]

Despite concerns about volatility, degradation, sensitivity to oxygen and micro-pollutants, and their overall high costs, there is a strong argument to be made that other technologies will be challenged to produce timely solutions for CO_2 control from conventional power plants.[31] The fact that amine scrubbing was patented in 1930 to remove CO_2 from natural gas and hydrogen,[34] and is being demonstrated industrially,[30] adds weight to these arguments. It remains an existential challenge for the development of any new CCS technology to be ready to be deployed at the industrial scale, given that experience has shown that bringing an energy technology from the bench to full-scale deployment may take at least three or more decades.[35]

2.3 Oxy-fuel Technology

Oxy-fuel technology is another major technology for CCS in which the fuel is burned in a nearly pure (90–95% pure O_2, depending on economics) oxygen stream to produce an off-gas consisting of CO_2 and steam, which can be condensed. The oxygen is produced by cryogenic separation of air, which is a well-established industrial-scale technology, but represents a significant energy penalty of around 5–10%.[36] In the primary embodiment of this technology, the fuel is burned in a pulverised coal-fired boiler.[37] In order to maintain boiler operation at normal temperatures approximately 70% of the flue gas must consist of recycled flue gas, and this is recycled usually after the water content has been reduced. With this technology about 90–95% of the CO_2 from the power plant is expected to be captured.[38]

Like amine scrubbing, there are no inherent technical restrictions on oxy-fuel technology and it has been demonstrated at the scale of 30 MWth (MWth: megawatt thermal which refers to the thermal power input to a plant), and a very large number of small pilot plants have been built to explore different aspects of the technology.[39] Currently, much of the discussion of such technologies is focused on cost, and various comparisons exist between pre- and post-combustion strategies and oxy-fuel among other technologies.[40–43] Currently, the various technical options for gas clean-up (*e.g.* nitrogen oxides: NO$_x$; sulfur oxides: SO$_x$; mercury: Hg; and other micro-pollutants) and the purity of CO_2 required sway the price comparisons significantly, but none of these technologies currently has an insurmountable cost advantage over the others. A key issue for this technology is that the boiler must minimise air leakage to something less than 3%.[36]

There are two other possible oxy-fuel options and these are circulating fluidised bed combustion (CFBC) oxy-fuel systems, and high-pressure oxy-fuel systems using natural gas. The first option has been demonstrated at the 30 MWth level at CIUDEN in Spain and appears to work well, with no insurmountable technical barriers.[44] The Spanish project also produced a

Front End Engineering Design (FEED) study for a 300 MWe unit.[45] The system employed was Foster Wheeler's Flexiburn™ technology which can operate in either air- or oxygen-fired mode and which appears to be fully fuel-flexible, thus representing a major technological step forward for CFBC technology.

CFBC in its oxy-fuel mode also offers the same advantages as in the case of air firing, which include lower NO_x emissions and inherent SO_2 capture if the bed is composed of limestone. The technology also offers the possibility of fuel flexibility, including biomass firing and hence neutral or negative CO_2 emissions (which will be discussed in Section 4). The one other major advantage of the CFBC option is that flue-gas recycle can be much less or, alternatively put, the oxygen levels introduced into the bed can be much higher, because some of the heat produced can be removed from the circulated solids by in-bed heat exchangers.

Another major possible route is the use of high-pressure oxy-turbine cycles with natural gas firing. This technology offers 100% capture and near-zero NO_x emissions with net efficiencies in the range of 43–65%, comparable to a combined-cycle gas turbine (CCGT) power plant, but with the benefits of CO_2 capture and lower emissions.[46] Despite the fact that there are many possible cycles being examined, this technology is still in its early stages, with only a few demonstration plants under development (most notably the NetPower, Texas and Clean Energy Systems or CES, California).[46] Nonetheless, given the importance of natural gas to many countries, this option has particular promise.

3   Chemical Looping and Calcium Looping Technologies

3.1   State of the Art Chemical Looping Technology

Chemical Looping Combustion (CLC) is arguably the newest of the major CCS technologies. It was initially patented as a method for producing high-purity CO_2 from hydrocarbons by Lewis and Gilliland,[47] following an earlier experimental demonstration by them of the oxidation of methane in a bed of copper oxide (CuO) particles.[48] Subsequently, it was explored by Richter and Knoche as a method of controlling the combustion process.[49]

At its heart it consists of two reactors: a fuel reactor where a hydrocarbon fuel is converted into either syngas or, in the case of combustion, CO_2 and H_2O, and an air reactor where the metal oxide is regenerated (see Figure 1). As with all such systems, the water can be condensed leaving a pure stream of CO_2 for use or storage.

Here, typical metal oxides explored as oxygen carriers are iron, copper, zinc, manganese and cobalt, and these metal oxides are used with different support materials.[50,51] Since oxygen carriers are solids the normal configuration envisaged is that of a dual fluidised bed. The first significant demonstration of CLC as a carbon capture technology was done at Chalmers University of Technology in Sweden in 2003 with a 10 kWth dual fluidised

N₂,O₂ CO₂,H₂O

MeO

Air reactor Fuel reactor

Me

Air Fuel

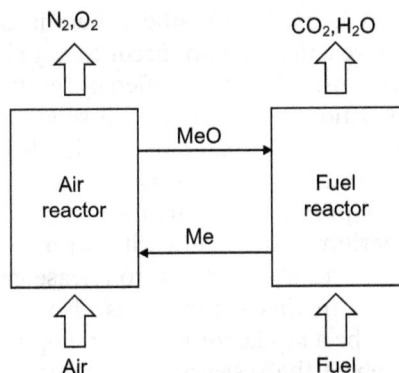

Figure 1 Chemical Looping Combustion (CLC) Reactor Configuration. (MeO = metal oxide, Me = metal).

bed prototype. Subsequently, there has been a vast range of papers and publications on different oxygen carrier-support combinations, and the pilot plants have increased in size to around 1 MWth. An examination using the search engine *Scopus* indicates the existence of over 1500 publications on the various aspects of chemical looping technology.

The critical advantage of this technology is that the heat balance in the two reactors is equivalent to direct conversion of the fuel in oxygen, as is exemplified below for the case of methane and the CuO system:

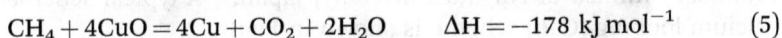

$$CH_4 + 4CuO = 4Cu + CO_2 + 2H_2O \qquad \Delta H = -178 \text{ kJ mol}^{-1} \qquad (5)$$

$$2Cu + O_2 = 2CuO \qquad \Delta H = -312.1 \text{ kJ mol}^{-1} \qquad (6)$$

Adding eqn (5) to 2× eqn (6) gives a heating value of $\Delta H = -802.2$ kJ mol^{-1} which is equivalent to burning the methane directly in oxygen but without the very high temperatures that would result or the need for cryogenic oxygen separation:

$$CH_4 + 2O_2 = CO_2 + 2H_2O \qquad \Delta H = -802.2 \text{ kJ mol}^{-1} \qquad (7)$$

Thus there is no need to supply heating or cooling between the fuel and air reactor, as is the case with calcium looping (CaL) for instance.

While gas combustion has been demonstrated with the standard gaseous hydrocarbon fuels, direct conversion of solid fuels is more problematic. However, in one of the most-recent costing studies, by Lyngfelt and Leckner,[52] these authors note that there has now been 2000 h of testing of solid-fuel CLC units and suggest that the additional cost of CLC-CFBC relative to CFBC technology is 20 € tCO₂$^{-1}$, with the largest part of this cost being that of CO₂ compression, a feature which is common to all capture technologies other than those where high-pressure oxygen is used in the first place.

Where the technology is used with solids, there is an inherent problem, namely that solid–solid reactions do not occur to any significant extent, and in that case the solid fuel must first be gasified by reaction with either CO_2 or H_2O, if the solid fuel is not first gasified in a separate step. One possible solution to this problem is to use materials which release gaseous oxygen at high temperatures. Interestingly, the release of oxygen from solids was one of the first methods of producing industrial oxygen in the BRIN process around 1884, where barium oxide reacts with air at 500 to 600 °C to form peroxide, which decomposes above 800 °C to release gaseous oxygen.[53] The possibility of using CuO in this manner was also explored by Lewis and Gilliland and is likely to be a key factor in the development of CLC with solid fuels.[47] Currently, much of the research on chemical looping with oxygen uncoupling (CLOU) materials is focussed on producing combined metal oxides, which have superior properties in terms of oxygen release. Ultimately, however, the largest challenge for CLC technology remains the need for large-scale demonstration units to allow this technology to compete with other commercial or near-commercial CCS options such as amine scrubbing and oxy-firing.

3.2 State of the Art Calcium Looping Technology

The idea of using lime to remove CO_2 from a syngas is relatively old.[54] However, its use in a cyclic scheme for CO_2 capture was first proposed by Professor Shimizu of Niigata University, Japan.[55] A typical schematic for a calcium looping (CaL) scheme is shown in Figure 2.

The basic concept is to use the reversible reaction:

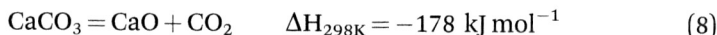

$$CaCO_3 = CaO + CO_2 \qquad \Delta H_{298K} = -178 \text{ kJ mol}^{-1} \qquad (8)$$

in a temperature-swing process with a high temperature driving the calcination reaction and a lower temperature being used for the carbonation process. This allows the transfer of CO_2 from one reactor to another, and permits the CO_2 to be concentrated to levels suitable for use or, more likely, for storage.[56,57] While temperature swing is the normal method considered,

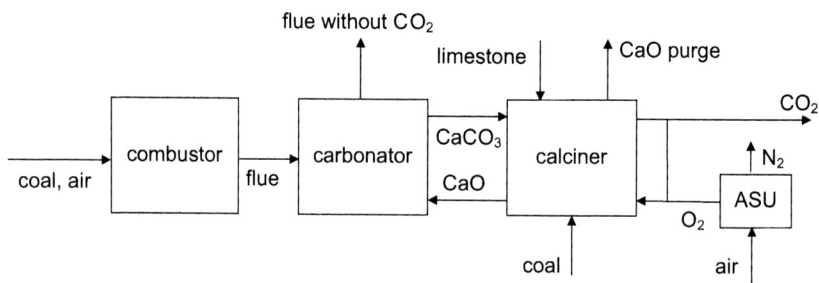

Figure 2 Calcium Looping (CaL) Schematic (ASU: Air Separation Unit). (Figure used with permission of Professor Paul Fennell, Imperial College, UK).

there has also been some work done on employing pressure swing to transfer the CO_2.[58]

The two major limitations of this technology are the fact that reactivity of calcium-based sorbents falls off very rapidly to reach levels of utilisation of about 10% or less by 20 reaction cycles, and the fact that sorbent loss by attrition can be considerable.[56] Nonetheless, studies support the idea that this technology is competitive when compared with amine scrubbing (with calculated energy penalties estimated at 6–8 percentage points, compared to 9.5–12.5% from amine-based post-combustion capture), principally because both the carbonator and the calciner can serve as part of the steam cycle when employed in the back-end CO_2 capture mode.[9] This represents an effective repowering of a boiler of around 30%.[56] Another possible advantage of the technology is that the spent lime from the process can be used in cement manufacture, where lime production represents some 50% of the CO_2 emissions.[57,59,60]

Like CLC, CaL has been extensively tested at the bench and pilot plant level[61] and there are two major 1–2 MWth demonstration units, which have shown that the technology is able to operate as expected with coal and natural limestones.[62–64] What remains, however, is the need for this technology to be tested at a much larger scale and, until this is done, this technology, like CLC, represents only a promising CCS option. In this context, it is interesting to note that both technologies have been assessed with a technical readiness level of 6 (where 1 represents an idea that is in concept only, and 10 an actual system has been demonstrated, while 6 means that there has been a prototype demonstration in a realistic environment).[65]

3.3   Hybrid Chemical and Calcium Looping Technologies

The idea of combining CaL and CLC (see Figure 3) first received attention as a possible fixed-bed process for producing hydrogen from the enhanced

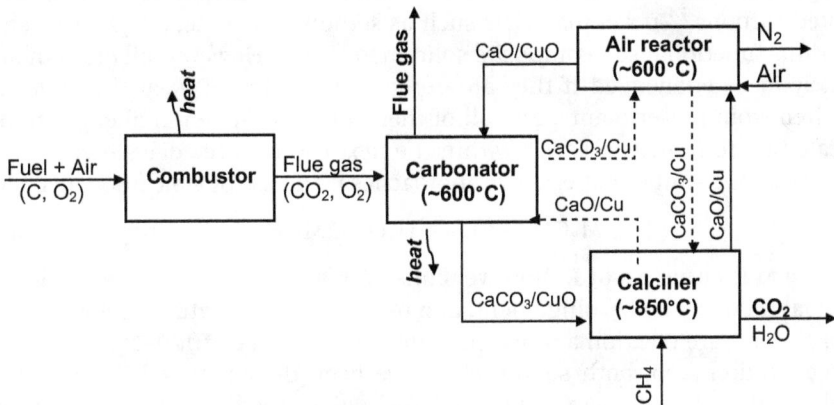

Figure 3   Alternative Scheme combining Calcium Looping (CaL) and Chemical Looping Combustion (CLC).

reforming process using natural gas.[66,67] Here the critical parameter is that Cu oxidation and reduction are both exothermic and, hence, can be used to drive the endothermic calcination reaction, thus eliminating the need for an oxy-fuel-fired calciner. Subsequently, efforts were made to produce pellets containing both copper and $CaCO_3$ which could operate in a fluidised bed environment and which could be used in a CO_2 capture cycle (see Figure 3).[68,69] Later, this idea was extended to the use of such composite sorbents for *in situ* CO_2 capture in a biomass gasifier; this work did, however, show some decline in the oxidation of the Cu, possibly due to $CaCO_3$ formation.[70] Fixed-bed tests also supported the idea that core-in-shell CaO/CuO-based pellets were particularly sensitive to deactivation and it was suggested that mixed pellets represented the best option.[71]

The idea of combining CaL and CLC has also been explored by Duhoux *et al.* using Aspen™ simulations.[72] The simulations showed that using a new solid looping configuration in the CaL-CLC process, to mitigate the loss of calcium oxide conversion after high cycle numbers, resulted in an improved process efficiency. The heat value was higher for the CaL-CLC process (34.2 *vs.* 31.2% for CaL alone) and the power output increased (by 136 *vs.* 110 MWe for CaL alone; for a 550 MWe power plant) due to the higher energy requirement to pre-heat the reactant. This is in line with the results of a study by Ozcana *et al.*[73] for a 250 MWe sub-critical power plant. Ozcana concluded that, compared to a 10.5% energy penalty for a conventional amine plant, the Ca–Cu looping process achieved the same overall CO_2 capture efficiency but with a significantly reduced energy penalty of 3.5 percentage points. This was the lowest energy penalty of the several processes examined (oxy-fuel, CaL and amine scrubbing).[73]

3.4   Alternative Solid CO_2 Capture Approaches

In principle, there are numerous possible solid CO_2 capture processes, besides CaL which has been discussed above. These include the use of activated carbons,[74] novel materials such as sodium zirconate, (Na_2ZrO_3)[75] and various functionalised composite solid materials.[76] However, all of them are likely to be challenged if they are expected to remove CO_2 at the scale required from power plants, and all of them will have major challenges to be scaled up to industrial levels within the next two or three decades.

One of the simplest alternative approaches is the use of a bicarbonate cycle:

$$M_2CO_3 + CO_2 + H_2O = 2MHCO_3 \qquad (9)$$

where M is either Na or K. Both versions of this cycle use particulate solid in a dual fluidised bed arrangement, and operate at temperatures below 100 °C for the capture reaction and temperatures in the range of 120–200 °C for the regeneration step; both such cycles have been demonstrated at the small pilot scale.[77] Like CaL and CLC, such technology requires demonstration at the industrial scale and, in the absence of such developments, is unlikely to be a major contributor to CCS technology.

4 Biomass with CO_2 Combustion (Bio-CCS)

4.1 Potential for Using Biomass in Negative CO_2 Schemes

Provided that biomass is converted in a sustainable manner (*i.e.*, grown at the same rate that it is harvested), it offers a simple method of CO_2 avoidance. Since biomass is probably the first fuel used by humanity, there is vast experience in combusting it and using it in all aspects of human society. A major limitation for terrestrial biomass is its low energy density, as it is difficult to justify transporting biomass over large distances unless it is first pelletised. Currently the market for pellets is around 10 million tonnes annually,[78] a figure that is dwarfed by coal production, which has increased from levels of about 3.4 to 7.9 billion tonnes from 1980 to 2013.[79] Providing enough biomass has often been an issue for co-firing plants, and there are new demands on biomass, such as the production of biofuels, which may further limit its future availability for direct combustion schemes. Nonetheless, providing it is available and can be fed to the device in question (which is particularly problematic for some kinds of biomass, such as straw), biomass can be burned directly and co-fired in both combustion and gasification systems. Work is currently underway to develop its use with CLC technology.[80]

4.2 Evaluation of Potential Bio-CCS Technology

In a major study on 28 bio-powered CCS technologies, Bhave and his colleagues concluded that while it was one of the few practical and economic means of removing large quantities of CO_2 from the atmosphere,[81] and the only approach that could also be used for electric power production, there was also a lack of financial incentives to do so. On the more positive side, this study concludes there are no major technical barriers to using biomass in CO_2-negative schemes. As stated previously, the major limits for this option will be providing enough biomass. In a study on the use of agricultural waste it was concluded that to provide the same amount of net electric energy such a plant must burn 30–90% more biomass than a co-fired plant, depending on the size and complexity of the plant.[82] In the absence of major financial initiatives, the full potential for biomass with CO_2 capture and storage of up to 10 Gt *p.a.*[83] is unlikely to be put into practice in the near future. Also, it is clear that conventional biomass by itself will not be sufficient to compensate for CO_2 production from fossil fuels (*ca.* 43 Gt *p.a.* of CO_2).

5 Air Capture, Mineralisation and CO_2 Utilisation (CCUS)

5.1 An Evaluation of Air Capture Options

The idea of removing CO_2 from the air is one that incites considerable controversy because, although air is relatively pure when compared with flue gas, the very low concentration of CO_2 of around 400 ppm implies the processing of very large amounts of gas as compared with flue gas, which might typically be composed of 12 to 15% CO_2. In a technical evaluation of direct air capture by the American Physical Society,[84] the costs of direct capture were estimated as being as high as \$600 ton^{-1} of CO_2. As a consequence, the argument was made that this technology should only be developed after all the significant point sources of CO_2 were eliminated, either by substitution of non-fossil fuel alternatives, or by capture of nearly all their CO_2 production. However, there are two situations in which the technologies that are being developed for air capture might look promising: first, as most CCS options offer only 90% CO_2 capture, such systems might conceivably be used as a polishing step for even deeper CO_2 reductions; and second, if after failing to meet all of our targets and abandoning fossil fuel use, we might still consider the global situation to be sufficiently dire to require remedial removal of CO_2 from the atmosphere.[84]

Air capture, as explored for instance by Keith and his collaborators,[85] involves the capture of atmospheric CO_2 by a caustic solution of either potassium or sodium hydroxide (KOH or NaOH), although numerous other systems such as the use of solid amines are possible.[86] The major challenge with this technology is the regeneration step, which must not be responsible for substantial CO_2 emissions to the atmosphere. Keith has also argued that air capture allows one to build such plants near the best storage sites and in regions where labour costs are low, and that air capture could also be used to match CO_2 emissions in situations where capture is not possible, such as aircraft or small combustion facilities.[87] While such arguments are compelling, the original objection that we must first fund CCS with major point sources, especially given the current world political situation where funding is not readily available, remains. In the current political climate it seems unlikely that air capture will become an important CCS technology in the near future.

5.2   A Primer on Mineralisation and its Potential

Mineralisation has a number of possible meanings for the CCS field. In one form it is the approach of capturing CO_2 using alkaline minerals (usually Ca-based) that are produced by various industrial processes, such as steel slag.[88] While this is an effective method for removing CO_2 from the atmosphere, the major issue remains that such minerals are available only in relatively small quantities (several hundred million tonnes *p.a.*[89] as compared to the 43 Gt *p.a.* of anthropogenic CO_2). Other possible routes include extraction of magnesium (Mg) from serpentine or similar minerals, but the major issue here is that the extraction step is likely to require significant processing of the minerals in question.[90] A major advantage of this approach, as opposed to using industrial wastes, is that Mg-containing minerals are ubiquitous and available in very large quantities (100 000 Gt).[91] Direct reaction of CO_2 with such minerals is unfortunately slow and requires high pressures and the use of significant grinding energy to prepare small enough particles to permit rapid reaction; it does not seem to be a very promising CCS option at this time. However, if the problems associated with either the extraction of Mg from such minerals or the direct reactions can be resolved, mineralisation offers one of the few solutions for capture of all of the anthropogenic CO_2 currently being produced.

5.3 CO_2 Utilisation and Enhanced Oil Recovery

Carbon capture and utilisation (CCU) is an attractive idea. However, there is a vigorous debate as to how much of the anthropogenic CO_2 production can be used in the manufacture of industrial chemicals or, perhaps, fuel. The primary limitation of such an approach is that the amounts of possible chemical products required are typically orders of magnitude (hundreds of millions of tonnes *p.a.*) less than the anthropogenic CO_2 production. Mac Dowell and co-workers have estimated the potential maximum use of CO_2 as a chemical feedstock as around 1%, with the best option for large CO_2 use in enhanced oil recovery being equivalent to utilisation of 4–8%. In their words: "from the perspective of mitigating climate change, CCU can at most be seen as supplementing CCS to a small extent.[92]

As noted above, enhanced oil recovery (EOR) represents the largest and most promising CO_2 utilisation technology, and also represents a revenue stream to drive a CCS project. Thus it is no accident that 7 of the 11 power plant projects listed on the MIT website were based on EOR.[13] It has also been argued that it represents a bridge to the future for carbon capture and storage, since it has been practised over many decades without problems. Despite this, the caveat still remains that EOR results in a net increase of carbon to the environment due to the desired production of fossil fuels. Indeed, in a 2010 study, US Department of Energy consultants concluded that there was "... little to suggest that CO_2-EOR is a necessary or significantly beneficial step towards the commercial deployment of CCS as a means of addressing climate change".[93]

6 Conclusions

CCS technology is currently available at industrial scales, with amine scrubbing and gasification being the main contenders for immediate large-scale deployment, and oxy-fuel firing as a close third. In addition, new technologies such as CLC and CaL are already demonstrated at the small pilot scale and could be fully commercialised over the next decade, given adequate funding. Given the lack of political will to support such projects, and the inability of renewables and good energy-storage technologies to offer us a carbon-constrained future in the next several decades, exceeding our CO_2 targets seems a more and more likely scenario. In that case, more extreme weather, forced migration and food shortages can be expected to become more severe, and mitigation options, which are the most expensive way of dealing with climate change, will become dominant.

Acknowledgements

The author would like to acknowledge Paul Fennell and Niall Mac Dowell at Imperial College and Stuart Haszeldine at the University of Edinburgh and Christopher Higman of Higman Consulting for useful discussions during the preparation of this chapter.

References

1. M. Hulme, *R. Meterol. Soc.*, 2009, 121.
2. S. Arrhenius, *Philos. Mag. J. Sci. Series*, 1896, **5**(41), 237.
3. S. Arrhenius, *Worlds in the Making*, Harper and Brothers Publishers, London and New York, 1907, p. 54.
4. P. N. Edwards, *A Vast Machine: Computer Models, Climate Data and the Politics of Global Warming*, MIT Press, USA, 2010, pp. 76–81.
5. J. Houghton, *Global Warming: The Complete Briefing*, Cambridge University Press, UK, 4th edn, 2009.
6. N. Stern, *The Economics of Climate Change: The Stern Review*, Cambridge University Press, UK, 2007, pp. 1–691.
7. N. Stern, *Nature*, 2016, **530**, 407.
8. J. C. J. M. Van Den Bergh, *Nat. Clim. Change*, 2017, **7**, 107.
9. M. E. Boot-Handford, J. C. Abanades, E. J. Anthony, M. J. Blunt, S. Brandani, N. Mac Dowell, J. R. Fernandez, M. C. Ferrari, R. Gross, J. P. Hallett, R. S. Haszeldine, P. Heptonstall, A. Lyngfelt, Z. Makuch, E. Mangano, R. T. J. Porter, M. Pourkashanian, G. T. Rochelle, N. Shah, J. G. Yao and P. S. Fennell, *Energy Environ. Sci.*, 2014, **7**, 130.
10. P. Collier, *The Bottom Billion: Why the Poorest Countries are Failing and What can be done about It*, Oxford University Press, UK, 2007.
11. J. M. Miocic, S. M. V. Gilfillan, J. J. Roberts, K. Elman, C. I. McDermott and R. Stuart Haszeldine, *Int. J. Greenhouse Gas Control*, 2016, **51**, 118.
12. *International Energy Outlook 2016*, Report No. DOE/EIA-0484(2016), released May 11, 2016.
13. *MIT: Carbon Capture and Sequestration Project Database, http://sequestration.mit.edu/tools/projects/*, (accessed March 2016).
14. H. Herzog, *Lessons Learned from CCS Demonstration and Large Pilot Projects*, An MIT Energy Initiative Working Paper, May 2016.
15. House of Commons Energy and Climate Change Committee: *Future of Capture and Storage in the UK, Second Report of Sessions 2015–2016*, 2 February 2016.
16. B. Fais, I. Keppo, M. Zeyringer, W. Usher and H. Daly, *Energy Strategy Rev.*, 2016, **13–14**, 154.
17. T. Vandyck, K. Keramidas, B. Savery, A. Kitous and Z. Vrontsi, *Global Environ. Change*, 2016, **41**, 46.
18. R. Gillard, *Global Environ. Change*, 2016, **40**, 26.
19. J. Tollefson, *Nat.: An Inte. Weekly J. Sci.*, 2015, DOI: 10.1038/nature.2015.16868.

20. R. Lofstedt, *J. Risk Res.*, 2015, **18**(6), 675.
21. S. Haszeldine, *Energy Environ.*, 2012, **23**, 437.
22. C. Higman and M. van der Burgt, *Gasification*, Gulf Professional Publishing, Houston, TX, USA, 2nd edn, 2008.
23. Gasification and Syngas Council, http://www.gasification-syngas.org/applications/gasification-overview/, (accessed March 2017).
24. C. Higman, personal communication, 2017.
25. V. M. Sikarwar, M. Zhao, P. Clough, J. Yao, Z. Zhong, M. Z. Memon, N. Shah, E. J. Anthony and P. S. Fennell, *Energy Environ. Sci.*, 2016, **9**, 2939.
26. L. Tan, Z. Dong, Z. Gong and M. Wang, *Renewable Energy*, 2017, **107**, 448.
27. F. Magelli, K. Boucher, H. T. Bi, S. Melin and A. Bonoli, *Biomass Bioenergy*, 2009, **33**, 434.
28. S. Gent, M. Twedt, C. Gerometta and E. Almberg, *Theoretical and Applied Aspects of Biomass Torrefaction: For Biofuels and Value-Added Products*, Elsevier, London, UK, 1st edn, 2017.
29. A. A. Thattai, V. Oldenbroek, L. Schoemakers, T. Woudstra and P. V. Aravind, *Appl. Thermal Eng.*, 2017, **14**, 170.
30. Boundary Dam Power Station, http://www.saskpower.com/our-power-future/our-electricity/our-electrical-system/boundary-dam-power-station/, (accessed March 2017).
31. G. T. Rochelle, *Science*, 2009, **325**, 1652.
32. Y. Du, Y. Y. Yuan and G. T. Rochelle, *Int. J. Greenhouse Gas Control*, 2017, **58**, 1.
33. A. B. Rao and E. S. Rubin, *Environ. Sci. Technol.*, 2002, **36**, 4467.
34. R. R. Bottoms, *US Pat.* 1 783 901A, 1930.
35. G. J. Kramer and M. Haigh, *Nature*, 2009, **462**, 568.
36. L. Zheng, *Oxy-fuel Combustion for Power Generation and Carbon Dioxide (CO$_2$) Capture*, Woodhouse Publishing, London, UK, 2011.
37. M. B. Toftegaard, J. Brix, P. A. Jensen, P. Glarborg and A. D. Jensen, *Prog. Energy Combust. Sci.*, 2010, **36**, 581.
38. M. Pehnt and J. Henkel, *Int. J. Greenhouse Gas Control*, 2009, **3**, 49.
39. R. Stanger, T. Wall, R. Spörl, M. Paneru, S. Grathwohl, M. Weidmann, G. Scheffknecht, D. McDonald, K. Myöhänen, J. Ritvanend, S. Rahiala, T. Hyppänen, A. Mletzkoe, A. Kather and S. Santos, *Int. J. Greenhouse Gas Control*, 2015, **40**, 55.
40. E. Rubin, J. E. Davison and H. J. Herzog, *Int. J. Greenhouse Gas Control*, 2015, **40**, 378.
41. R. T. J. Porter, M. Fairweather, C. Kolster, N. Mac Dowell, N. Shah and R. M. Woolley, *Int. J. Greenhouse Gas Control*, 2017, **57**, 185.
42. X. D. Wu, Q. Yang, G. Q. Chen, T. Hayat and A. Alsaedi, *Renewable Sustainable Energy Rev.*, 2016, **60**, 1274.
43. K. Atsonios, K. Panopoulos, P. Grammelis and E. Karakas, *Int. J. Greenhouse Gas Control*, 2016, **45**, 106.
44. E. J. Anthony and H. Hack, in *Fluidized Bed Technologies for Near-Zero Emissions Combustion and Gasification*, ed. F. Scala, Woodhouse Publishing, London, UK, 2013.

45. OxyCFB300, https://hub.globalccsinstitute.com/sites/default/files/publications/137158/Compostilla-project-OXYCFB300-carbon-capture-storage-demonstration-project-knowledge-sharing-FEED-report.pdf (accessed March 2017).

46. F. C. Barba, G. M. Sánchez, B. S. Seguí, H. G. Darabkhani and E. J. Anthony, *J. Cleaner Prod.*, 2016, **133**, 971.

47. W. K. Lewis and E. R. Gilliland, *US Pat.* 2 665 972, 1954.

48. W. K. Lewis, E. R. Gilliland and W. A. Reed, *Ind. Eng. Chem.*, 1949, **41**, 1227.

49. H. J. Richter and K. F. Knoche, *ACS Symp. Ser.*, 1983, 71.

50. L. S. Fan, *Chemical Looping Systems for Fossil Fuel Energy Conversion*, Wiley, New Jersey, USA, 2010.

51. J. Adanez, A. Abad, F. Garcia-Labiano, P. Gayan and L. F. de Diego, *Prog. Energy Combust. Sci.*, 2012, **38**, 215.

52. A. Lyngfelt and B. Leckner, *Appl. Energy*, 2015, **157**, 475.

53. BRIN, https://en.wikipedia.org/wiki/Brin_process, (accessed March 2017).

54. Tessié du Motay et Maréchal, Bull Mensuel de la Société Chimique de Paris, 1868.

55. T. Shimizu, T. Hirama, H. Hosoda, K. Kitano, K. M. Inagaki and K. Teijima, *Chem. Eng. Res. Des.*, 1999, **77**, 62.

56. J. Blamey, E. J. Anthony, J. Wang and P. S. Fennell, *Prog. Energy Combust. Sci.*, 2010, **36**, 260.

57. *Calcium and Chemical Looping Technology for Power Generation and Carbon Dioxide (CO_2) Capture*, ed. P. Fennell and B. Anthony, Woodhead Publishing, London, UK, 2015.

58. J. W. Butler, C. Jim Lim and J. R. Grace, *Fuel*, 2014, **127**, 78.

59. C. C. Dean, J. Blamey, N. H. Florin, M. J. Al-Jeboori and P. S. Fennell, *Chem. Eng. Res. Des.*, 2011, **89**, 836.

60. A. Telesca, M. Marroccoli, M. Tomasulo, G. Lorenzo Valenti, H. Dieter and F. Montagnaro, *Environ. Sci. Technol.*, 2015, **49**, 6865.

61. D. P. Hanak, E. J. Anthony and V. Manovic, *Energy Environ. Sci.*, 2015, **8**, 2199.

62. J. Ströhle, M. Junk, J. Kremer, A. Galloy and B. Epple, *Fuel*, 2014, **127**, 13.

63. B. Arias, M. E. Diego, J. C. Abanades, M. Lorenzo, L. Diaz, D. Martínez, J. Alvarez and A. Sánchez-Biezma, *Int. J. Greenhouse Gas Control*, 2013, **18**, 237.

64. M. E. Diego, B. Arias, A. Méndez, M. Lorenzo, L. Diaz, A. Sánchez-Biezma and J. C. Abanades, *Int. J. Greenhouse Gas Control*, 2016, **50**, 14.

65. J. C. Abanades, B. Arias, A. Lyngfelt, T. Mattisson, D. E. Wiley, H. Li, M. T. Hoc, E. Mangano and S. Brandani, *Int. J. Greenhouse Gas Control*, 2015, **40**, 126.

66. J. R. Fernández, J. C. Abanades, R. Murillo and G. Grasa, *Int. J. Greenhouse Gas Control*, 2012, **6**, 126.

67. I. Martínez, M. C. Romano, J. R. Fernández, P. Chiesa, R. Murillo and J. C. Abanades, *Appl. Energy*, 2014, **114**, 192.

68. V. Manovic and E. J. Anthony, *Environ. Sci. Technol.*, 2011, **45**, 10750.
69. V. Manovic and E. J. Anthony, *Energy Fuels*, 2011, **25**, 4846.
70. R. A. Rahman, P. Mehrani, D. Y. Lu, E. J. Anthony and A. Macchi, *Energy Fuels*, 2015, **29**, 3808.
71. F. N. Ridha, D. Lu, A. Macchi and R. W. Hughes, *Fuel*, 2015, **153**, 202.
72. B. Duhoux, P. Meharani, D. Y. Lu, R. T. Symonds, E. J. Anthony and A. Macchi, *Energy Technol.*, 2016, **4**, 1158.
73. D. C. Ozcana, A. Macchi, D. Y. Lu, A. M. Kierzkowska, H. Ahna, C. R. Müller and S. Brandani, *Int. J. Greenhouse Gas Control*, 2015, **43**, 198.
74. K. Singh, I. Y. Kim, K. S. Lakhi, P. Srivastava, R. Naidu and A. Vinue, *Carbon*, 2017, **116**, 448.
75. G. Ji, M. Z. Memon, H. Zhuo and M. Zhao, *Chem. Eng. J.*, 2017, **313**, 646.
76. A. L. Yaumi, M. Z. Abu Bakar and B. H. Hameed, *Energy*, 2017, **124**, 461.
77. C. Zhao, X. Chen, E. J. Anthony, X. Jiang, L. Duan, Y. Wu, W. Dong and C. Zhao, *Prog. Energy Combust. Sci.*, 2013, **39**, 515.
78. C. Whittaker and I. Shield, *Renewable Sustainable Energy Rev.*, 2017, **71**, 1.
79. IndexMundi, http://www.indexmundi.com/energy/?product = coal, (accessed February 2017).
80. B. Vilches, F. Lind, M. Rydén and H. Thunman, *Appl. Energy*, 2017, **190**, 1174.
81. A. Bhave, R. H. S. Taylor, P. Fennell, W. R. Livingston, N. Shah, N. Mac Dowell, J. Dennis, M. Kraft, E. M. Pourkashanian, M. Insa, J. Jones, N. Burdett, A. Bauen, C. Beal, A. Smallbone and J. Akroyd, *Appl. Energy*, 2017, **190**, 481.
82. J. Hetland, P. Yowargana, S. Leduc and F. Kraxner, *Int. J. Greenhouse Gas Control*, 2016, **59**, 330.
83. International Energy, *Potential for Biomass and Carbon Dioxide Capture and Storage*, Report 2011/06, July 2011.
84. APS, *Direct Air Capture of CO_2 with Chemicals: A Technology Assessment for the APS Panel on Public Affairs*, June 1, 2011.
85. M. Mahmoudkhani and D. W. Keith, *Int. J. Greenhouse Gas Control*, 2009, **3**, 376.
86. L. A. Darunte, A. D. Oetomo, K. S. Walton, D. S. Sholl and C. W. Jones, *Sustainable Chem. Eng.*, 2016, **4**, 5761.
87. D. W. Keith, *Science*, 2009, **25**, 1654.
88. A. Polettini, R. Pomi and A. Stramazzo, *J. Environ. Manage.*, 2016, **167**, 185.
89. U.S. Geological Survey, *Mineral Commodities Summaries*, February 2014.
90. A. Sanna, L. Steel and M. M. Maroto-Valer, *J. Environ. Manage.*, 2017, **189**, 84.
91. K. S. Lackner, *Annu. Rev. Energy Env.*, 2002, **27**, 190.
92. N. Mac Dowell, P. S. Fennell, N. Shah and G. C. Maitland, *Nat. Climate Change*, 2017, **7**, 243.
93. J. J. Dooley, R. T. Dahowski and C. L. Davidson, *CO_2-driven Enhanced Oil Recovery as a Stepping Stone to What?*, USDoE under Contract DE-AC05-76RL01830, July 2010.

Subject Index

www.ingramcontent.com/pod-product-compliance
Lightning Source LLC
Chambersburg PA
CBHW031949180326
41458CB00006B/1670